TUMOR MICROENVIRONMENT REGULATION OF TUMOR EXPANSION

TUMOR MICROENVIRONMENT REGULATION OF TUMOR EXPANSION

DOMENICO RIBATTI

Human Anatomy, Department of Basic Biomedical Sciences, Neurosciences and Sensory Organs, Section of Human Anatomy and Histology, University of Bari Medical School, Bari, Italy

ELSEVIER

ACADEMIC PRESS

An imprint of Elsevier

Academic Press is an imprint of Elsevier
125 London Wall, London EC2Y 5AS, United Kingdom
525 B Street, Suite 1650, San Diego, CA 92101, United States
50 Hampshire Street, 5th Floor, Cambridge, MA 02139, United States
The Boulevard, Langford Lane, Kidlington, Oxford OX5 1GB, United Kingdom

Library of Congress Cataloging-in-Publication Data
A catalog record for this book is available from the Library of Congress

British Library Cataloguing-in-Publication Data
A catalogue record for this book is available from the British Library

ISBN: 978-0-12-822803-6

For information on all Academic Press publications visit our website at https://www.elsevier.com/books-and-journals

Publisher: Stacy Masucci
Acquisitions Editor: Rafael E. Teixeira
Editorial Project Manager: Ruby Smith
Production Project Manager: Selvaraj Raviraj
Cover Designer: Miles Hitchen

Typeset by TNQ Technologies

Contents

Preface

The biological relationship between the living being and its environment is a functional relationship and, as such, dynamic. In the light of these considerations, individual phenomena cannot be considered as exemplifications of laws of nature considered as invariable, and therefore as such they do not come close to the integrality of a supposed normative reality and, consequently, they are not able to reproduce it adequately. Ultimately, it is the environment that decides the success of a species or its appearance.

It is the environment that makes a selection between the different species and between the various populations and individuals of the same species. As the environment changes constantly, the characteristics of living things also change over time. Individuals in a certain environment learn to make better use of it. The changes that occur over the generations shape the adaptation, both by expanding the ability to exploit the environment in which they find themselves and by deviating from their own pattern to adapt to an environment other than the primitive.

The environment acts by selecting the individuals who correspond to it, favoring some over others: individuals who, by chance, find themselves better adapted to their surroundings develop better and reproduce more.

Carcinogenesis is a multiphase process in which the modification of the characteristics of the extracellular microenvironment is necessary for the transformation of a normal cell into a tumor cell. The importance of the microenvironment in promoting and supporting neoplastic growth is widely recognized, and considering the microenvironment as an elective target for a series of new therapeutic approaches is an established concept in oncology. The microenvironment is defined as the set of normal cells with which transformed cells live and from which they draw nourishment and proliferative and antiapoptotic stimuli. All tumors have a stroma that consists of connective tissue, blood vessels, hematopoietic cells, stromal cells, fibroblasts, and an inflammatory infiltrate, which can make up as much as 90% of the tumor mass.

The extracellular matrix that enters the composition of the stroma is made up of a set of proteins such as collagen, glycoproteins, and proteoglycans, whose composition and relative abundance determines its architectural structure and physicochemical characteristics.

The microenvironment of cancer cells is important in determining their phenotypic instability. The close contact between tumor and stroma exposes the stromal cells to a variety of proteolytic enzymes secreted by tumor cells, which modify the composition of the extracellular matrix favoring the invasiveness of the tumor cells themselves and the release of growth factors. Changes in the extracellular matrix surrounding a tumor cell population can lead to alterations in their gene expression and cellular phenotype. The reconstitution of a normal microenvironment can lead to a tumor growth arrest with reversion of the malignant phenotype.

The tumor microenvironment affects the therapeutic response and resistance of the tumor, which is why new therapies are being developed to target specific components of the environment. Much research and development are being conducted to further understand the relationship between different kinds of tumors and the tumor microenvironment to establish effective new treatments.

Domenico Ribatti

CHAPTER 1

Tumor microenvironment

1.1 Introduction

Cancer remains a leading cause of mortality worldwide. In the United States, over 1.5 million of new cases were diagnosed in 2016, leading to 595,000 deaths (Siegel, Miller, & Jemal, 2016). Doug Hanahan and Robert A. Weinberg (2000, 2011) postulated six acquired capabilities, or hallmarks of cancer, that are shared by most human tumors: self-sufficiency in growth signals (tumors have the capacity to proliferate without external stimuli due to activation of oncogenes), insensitivity to antigrowth signals (tumors may not respond to molecules that inhibit the proliferation of normal cells due to inactivation of tumor suppressor genes), evasion of apoptosis (tumors are resistant to programmed cell death), limitless replicative potential (tumors have unrestricted proliferative capacity, a stem cell-like property), sustained angiogenesis, and tissue invasion and metastasis (Fig. 1.1). Together, all these capabilities allow cancer cells to survive, proliferate, and disseminate. In 2011, the hallmarks of cancer were revisited and extended. Genomic instability, reprogramming of energy metabolism, tumor-induced inflammation, and escape from immune destruction are recognized as additional hallmarks that contribute and foster tumor development and progression (Fig. 1.1).

Interactions with the tumor stroma were also highlighted to contribute to the acquirement of hallmark traits. The final emerging cancer hallmark is metabolic reprogramming of tumor cells. By means of Warburg effect, tumor cells shift from predominantly aerobic glycolysis toward anaerobic glycolysis (Note 1). Even in the presence of high oxygen tension, tumor cells exploit the anaerobic metabolic process to facilitate rapid cell division (Gillies, Robey, & Gatenby, 2008). Cancer research until the 1980s was dominated by a tumor-centric view postulating that mutations in oncogenes and tumor suppressor genes were sufficient to determine carcinogenesis and cancer progression (Note 2) (Vogelstein & Kinzler, 1993). The

Tumor Microenvironment Regulation of Tumor Expansion
ISBN 978-0-12-822803-6
https://doi.org/10.1016/B978-0-12-822803-6.00008-9

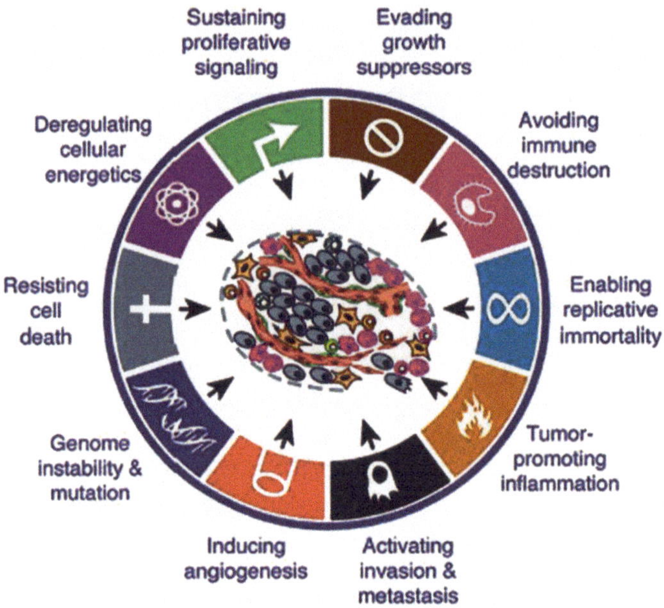

Figure 1.1 The hallmarks of cancer. *(Modified from Hanahan, D., & Weinberg, R. A. (2011). Hallmarks of cancer: The next generation. Cell, 144(5), 646–674. https://doi.org/10.1016/j.cell.2011.02.013.)*

prevailing paradigm for the development of cancer is a multistep process during which a cell acquires multiple genetic mutations (Weinberg, 1989). Genetic alterations in the tumor cells themselves are not sufficient to generate a malignant tumor but that a permissive stromal environment is needed as well involved coordinated actions of stromal components like inflammatory cells, blood vessels, and extracellular matrix (ECM).

1.2 Tumor microenvironment

Tumor mass dormancy or angiogenic dormancy occur when proliferation is balanced by apoptosis because of a lack of vasculature and limited supply of nutrients and oxygen. Reestablishment of vasculature is favored by hematopoietic precursor cells (HPCs) and endothelial precursor cells (EPCs) expressing vascular endothelial growth factor receptors (VEGFRs) and by dendritic cell precursors that can differentiate into endothelial-like cells. Once tumor cells overcome dormancy, they become receptive to other cells which further support their expression.

The tissue microenvironment undergoes extensive remodeling during morphogenesis and tumor progression, including changes in the deposition, degradation, and structural organization of the ECM components. The remodeling of the ECM is involved in the control of cell survival, proliferation, migration, and differentiation (Nelson & Bissell, 2006). In the 1980s, literature data have shown that tumor microenvironment–derived signals delivered to cancer cells reprogramme the phenotype of tumor cells and determine their survival and progression. Tumor microenvironment plays an important role in the initiation and progression of tumors (Balkwill & Mantovani, 2001; Hussain & Harris, 2007). Generally, the microenvironment of early-stage tumors tends to exert antimalignancy functions, whereas that of late-stage tumors tends to exert promalignancy functions (Klein-Goldberg, Maman, & Witz, 2014).

Studies on neoplastic transformation have focused on events that occur within transformed cells and have addressed the microenvironment of tumor cells documenting its importance in supporting tumor progression. The pathogenesis of most cancers includes complex and mutual interactions that affect the number and phenotype of the tumor cells and various normal stromal cells, and intricate tumor microenvironment interactions are increasingly recognized as critical features of several tumors. Laplane, Duluc, Larmonier, Pradeu, and Bikfalvi (2018) have recently raised some important semantic questions correlated to the definition of tumor microenvironment: "Is the tumor microenvironment the pretumoral site that favors the development of a tumor, or the local environment induced by the tumor? (Laconi, 2007). (...) Some researches consider that the tumor is a part of the tumor microenvironment, while others understand the tumor microenvironment as all the non-tumor components surrounding the tumor (Li, Fan, & Houghton, 2007)." Stromal tissue is much more than a passive bystander in the development and progression of cancer. Instead, there is a bidirectional, dynamic, and intricate complex of interactions between the stromal and cancer cells, in which tumor cells contribute to the generation and modification of tumor microenvironment. In this context, tumor microenvironment is in constant evolution as a result of tissue remodeling, metabolic alterations in the tumor, and changes in the recruitment of stromal cells. For example, human mammary tumor cells implanted into mouse mammary fat pads formed tumors more readily, with increased metastatic potential, when implanted in involuting (postlactation) mammary glands, rather than during pregnancy (Borges & Schedin, 2012).

1.3 Tumor hypoxia and interstitial fluid pressure

In 1955, Thomlinson and Gray (1955) analyzing histological sections of carcinomas observed that tumor regions close to blood vessels were viable, whereas regions in the center of the tumor were necrotic, expression of the existence of hypoxia in the tumor core. According to Laplane et al. (2018), a metabolic symbiosis has been described between hypoxic and oxygenated tumor cells. Metabolic adaptation of the tumor cells to intratumoral hypoxia includes switching from oxidative to glycolytic pathways. Tumors are often hypoxic in spite of high vascularization due to the poor structure and functionality of tumor blood vessels.

Tumor vessels are immature and may be dilated, tortuous, and hyperpermeable due an often incomplete or absent basement membrane, and they are prone to excessive branching and neovascular shunts. Poorly connected endothelial cells, loosely associated vascular smooth muscle cells, and an increase in vesiculo-vascuolar organelles contribute to this hyperpermeable phenotype. Over 50 per cent of locally advanced tumors present hypoxic and even anoxic areas, with extreme heterogeneity within the tumor mass (Semenza, 2012). In fact, the extent of hypoxic regions is heterogeneous even among tumors of identical histopathological type and does not correlate with tumor size, stage, and grade (Hockel & Vaupel, 2001). Hypoxia in tumors develops in the form of chronic hypoxia, resulting from long diffusion distances between perfused tumor vessels and/or of acute hypoxia, resulting from a transient collapse of tumor vessels. Hypoxia inducible factors (HIFs) are activated in these series of events (Note 3). Intratumoral hypoxia is a main cause of high reactive oxygen species (ROS) formation within the tumor cells (Guzy & Schumacker, 2006) and also is coupled to pathological tumor cell metabolism and acidosis (Denko, 2008). Hypoxic tumor cells have the ability to create an immunosuppressive microenvironment. Through the ability to produce HIF-1 alpha (HIF-1α), tumor necrosis factor alpha (TNF-α), and stromal cell derived factor-1 (SDF-1), hypoxic tumor cells produce the homing of $CD45^+$ myeloid cells to tumor site, including subpopulations defined by the expression of Tie-2, VEGFR-1, and/or CD11b, as well as tumor associated macrophages (TAMs) (Du et al., 2008; Mantovani, Allavena, Sica, & Balkwill, 2008).

Stimulation of angiogenesis in the context of ongoing inflammation is usually related to hypoxia. HIF—NF-kB cross talk present a basal molecular mechanism influencing inflammatory angiogenesis. NF-kB is a direct modulator of HIF-1α expression, as NF-kB binding site is present in HIF-1α promoter region (Table 1.1). Hypoxia upregulates various angiogenic factors, including VEGF (Note 4), angiopoietin-2 (Ang-2), platelet derived growth factor (PDGF), transforming growth factor alpha (TGFα), interleukin-8 (IL-8), and hepatocyte growth factor (HGF) (Harris, 2002). HIF1α, the master regulator of oxygen homeostasis (Semenza, 2003), binds to the hypoxia-responsive element (HRE) in the promoter of hypoxia-responsive genes, such as VEGF, PDGF, and TGFα, and induces their expression (Harris, 2002; Semenza, 2003). Other factors, such as IL-8 and placental growth factor (PlGF), are activated by HIF-independent mechanisms (Harris, 2002). Hypoxia mediates immune cell recruitment, and these cells concentrate at the tumor periphery, while in the tumor core hypoxia contributes to cancer cell escape by providing an aggressive selection for stem-like tumor cells that migrate away to the tumor margin. Hypoxia generates an immunosuppressive microenvironment, which inhibits the effects of cytotoxic $CD8^+$ cells and polarizes TAMs into the proangiogenic M2 phenotype. Hypoxic areas are refractory to chemo- and radiotherapy and contribute to select tumor populations able to survive in poorly oxygenated niches and escape to metastatic sites and proangiogenic cancer stem cells (CSCs). HIF-1α acts as a survival factor for cancer cells by activating the transcription of genes involved in angiogenesis, proliferation, and migration of endothelial cells, pericyte recruitment, and changes in vascular permeability (Table 1.2). Persistent extravasation of cells enables activation of angiogenic response due to the constant and abundant release of inflammatory stimuli, which are also proangiogenic molecules or stimulate generation of such factors. Hypoxia can increase the invasive potential

Table 1.1 Proangiogenic factors induced by hypoxia via NF-kB pathway.

Interleukin-1 beta, Interleukin-6, Interleukin-8 (IL-1β, IL-6, IL-8)
Cyclooxygenase-2 (COX-2)
Tumor necrosis factor alpha (TNF-α)
Macrophage inflammatory protein-2 (MIP-2)
Intercellular adhesion molecule-1 (ICAM-1)
Vascular cell adhesion molecule-1 (VCAM-1)
Chemokine (C—C motif) ligand-5 (CCL-5)
Inducible nitric oxide synthases (iNOS)

Table 1.2 Genes activated by hypoxia inducible factor-1 alpha.

Involved in angiogenesis
Vascular endothelial growth factor (VEGF)
Placental growth factor (PlGF)
Vascular endothelial growth factor receptor-1 (VEGFR-1)
Involved in endothelial cell proliferation and migration
Vascular endothelial growth factor (VEGF)
Fibroblast growth factor-2 (FGF-2)
C-X-C motif chemokine
Ligand 12/C-X-C chemokine receptor type 4 (CXCL12/CXCR4)
Platelet derived growth factor (PDGF)
Involved in pericyte recruitment
Platelet derived growth factor (PDGF)
Angiopoietin-1 (Ang-1)
Involved in changes in vascular permeability
Vascular endothelial growth factor (VEGF)
Vascular endothelial growth factor receptor-1 (VEGFR-1)
Angiopoietin-1 (Ang-1)

of cancer cells by inducing the production of promigratory proteins (e.g., SDF-1α and HGF) and proinvasive extracellular matrix proteins (Semenza, 2014). HIF-1α and HIF-2 promote the expression of other transcription factors, including Twist, Snail-1, Zeb-1, BMI-1, and Notch. Twist-1 is associated with the acquisition of stem cell properties and enhancement of metastasis (Yang et al., 2004). Another important feature of the tumor microenvironment is the elevated interstitial fluid pressure (IFP) with values ranging from 10 to 100 mmHg (Note 5) (Less et al., 1992).

The highly irregular tumor neovasculature is characterized by increased permeability and microvascular resistance, which results in the movement of fluid into the ECM. As tumors are lacking in lymphatic vessels, excess fluid accumulates in the interstitium elevating IFP. In addition, as cancer and stromal cells proliferate and the tumor mass grows larger, the mechanical stress obliterates the vascular lumen (Brú, Albertos, Luis Subiza, García-Asenjo, & Brú, 2003). Elevated IFP is a barrier to delivery of therapeutic agents reducing their efficacy, disrupting the development of an optimal gradient, and hindering drug delivery to the tumor core (Boucher, Baxter, & Jain, 1990). Improved uptake of chemotherapeutic agents occurs following a reduction in IFP. Bissell and Radisky (2001) pointed out that tumors evolve from an organ and retain memory of that organ and that once they liberate themselves from the constraints of the normal tissue microenvironment and lose the organ-specific structure as they evolve into the tumor organs and constantly redefine themselves by their ever-changing context (Du et al., 2008; Mantovani et al., 2008).

1.4 Notes

1. In the 1920s, Otto Warburg and colleagues discovered that tumors, unlike normal cells, converted glucose to pyruvate and then to lactate, even in the presence of plentiful amount of oxygen (the "Warburg effect"). They made the observation that tumors were taking up enormous amounts of glucose compared to what was seen in the surrounding tissue. Additionally, glucose was fermented to produce lactate even in the presence of oxygen, thus the term aerobic glycolysis. Warburg received the Nobel Prize in 1931 for discovery of the effect that bears his name. The Warburg effect has been documented for over 90 years and extensively studied over the past 10 years with thousands of papers reporting to have established either its causes or its functions. Despite this intense interest, the function of the Warburg effect remains unclear. Warburg's discovery was largely neglected for many years, but over the past decade metabolism has become one of the most active areas of cancer research.
2. Tumor growth and progression occurs as a result of cumulative acquisition of genetic alterations affecting oncogenes or tumor suppressor genes. Proto-oncogenes are genes that normally help cells grow. When a proto-oncogene mutates or there are too many copies of it, it becomes a "bad" gene that can become permanently turned on or activated (oncogene). When this happens, the cell grows out of control, which can lead to cancer. Tumor suppressor genes are normal genes that slow down cell division, repair DNA mistakes, or tell cells when to die (apoptosis or programmed cell death). When tumor suppressor genes do not work properly, cells can grow out of control, which can lead to cancer. Proangiogenic proteins including VEGF are upregulated by oncogenes, while antiangiogenic proteins are downregulated (Rak, Yu, Klement, & Kerbel, 2000).
3. Hypoxia-inducible factor (HIF) is a sequence-specific DNA-binding protein that can promote or repress the transcription of a broad range

of genes that are involved in maintaining biological homeostasis. HIF is a heterodimeric complex that is composed of an oxygen-destructible α-subunit and an oxygen-indestructible β-subunit. Three isoforms of the α-subunit and two isoforms of the β-subunit are thought to be involved in the in vivo response to hypoxia. HIF is mostly nonfunctional in oxygenated cells but becomes active under specific conditions, including low-oxygen (hypoxic) stress. HIF mediates coordinate upregulation of genes of the glycolytic pathway ranging from glucose uptake to lactate production. Lactate and pyruvate, that accumulate as a consequence of the "Warburg effect" further promote HIF stabilization and gene expression (Lu, Forbes, & Verma, 2002).

4. The mammalian VEGF family comprises five mail ligands (VEGF-A, VEGF-B, VEGF-C, and VEGF-D, and placental growth factor (PlGF) and three receptors (VEGFR-1, VEGF-2, and VEGF-3). The principal role of VEGF-C and VEGF-D is to stimulate lymphangiogenesis through VEGFR-3, while blood vessel growth is stimulated by VEGF-A through VEGFR-1 and VEGFR-2.
5. Interstitial fluid pressure is a physiological parameter with demonstrated predictive value for a tumor's aggressiveness, drug delivery, as well as response to treatments such as radiotherapy and chemotherapy. However, its clinical use has been limited because the technique available for its measurement has been invasive.

References

Balkwill, F., & Mantovani, A. (2001). Inflammation and cancer: Back to Virchow? *Lancet, 357*(9255), 539–545. https://doi.org/10.1016/S0140-6736(00)04046-0.

Bissell, M. J., & Radisky, D. (2001). Putting tumours in context. *Nature Reviews Cancer, 1*(1), 46–54. https://doi.org/10.1038/35094059.

Borges, V. F., & Schedin, P. J. (2012). Pregnancy-associated breast cancer: An entity needing refinement of the definition. *Cancer, 118*(13), 3226–3228. https://doi.org/10.1002/cncr.26643.

Boucher, Y., Baxter, L. T., & Jain, R. K. (1990). Interstitial pressure gradients in tissue-isolated and subcutaneous tumors: Implications for therapy. *Cancer Research, 50*(15), 4478–4484.

Brú, A., Albertos, S., Luis Subiza, J., García-Asenjo, J. L., & Brú, I. (2003). The universal dynamics of tumor growth. *Biophysical Journal, 85*(5), 2948–2961. https://doi.org/10.1016/S0006-3495(03)74715-8.

Denko, N. C. (2008). Hypoxia, HIF1 and glucose metabolism in the solid tumour. *Nature Reviews Cancer, 8*(9), 705–713. https://doi.org/10.1038/nrc2468.

Du, R., Lu, K. V., Petritsch, C., Liu, P., Ganss, R., Passegué, E., et al. (2008). HIF1alpha induces the recruitment of bone marrow-derived vascular modulatory cells to regulate tumor angiogenesis and invasion. *Cancer Cell, 13*(3), 206–220. https://doi.org/10.1016/j.ccr.2008.01.034.

Gillies, R. J., Robey, I., & Gatenby, R. A. (2008). Causes and consequences of increased glucose metabolism of cancers. *Journal of Nuclear Medicine, 49*(6). https://doi.org/10.2967/jnumed.107.047258.

Guzy, R. D., & Schumacker, P. T. (2006). Oxygen sensing by mitochondria at complex III: The paradox of increased reactive oxygen species during hypoxia. *Experimental Physiology, 91*(Issue 5), 807–819. https://doi.org/10.1113/expphysiol.2006.033506. Blackwell Publishing Ltd.

Hanahan, D., & Weinberg, R. A. (2000). The hallmarks of cancer. *Cell, 100*(1), 57–70. https://doi.org/10.1016/S0092-8674(00)81683-9.

Hanahan, D., & Weinberg, R. A. (2011). Hallmarks of cancer: The next generation. *Cell, 144*(5), 646–674. https://doi.org/10.1016/j.cell.2011.02.013.

Harris, A. L. (2002). Hypoxia — a key regulatory factor in tumour growth. *Nature Reviews Cancer, 2*(1), 38–47. https://doi.org/10.1038/nrc704.

Hockel, M., & Vaupel, P. (2001). Tumor hypoxia: Definitions and current clinical, biologic, and molecular aspects. *Journal of the National Cancer Institute, 93*(4), 266–276. https://doi.org/10.1093/jnci/93.4.266.

Hussain, S. P., & Harris, C. C. (2007). Inflammation and cancer: An ancient link with novel potentials. *International Journal of Cancer, 121*(11), 2373–2380. https://doi.org/10.1002/ijc.23173.

Klein-Goldberg, A., Maman, S., & Witz, I. P. (2014). The role played by the microenvironment in site-specific metastasis. *Cancer Letters, 352*(1), 54–58. https://doi.org/10.1016/j.canlet.2013.08.029.

Laconi, E. (2007). The evolving concept of tumor microenvironments. *BioEssays, 29*(8), 738–744. https://doi.org/10.1002/bies.20606.

Laplane, L., Duluc, D., Larmonier, N., Pradeu, T., & Bikfalvi, A. (2018). The multiple layers of the tumor environment. *Trends in Cancer, 4*(12), 802–809. https://doi.org/10.1016/j.trecan.2018.10.002.

Less, J. R., Posner, M. C., Boucher, Y., Borochovitz, D., Wolmark, N., & Jain, R. K. (1992). Interstitial hypertension in human breast and colorectal tumors. *Cancer Research, 52*(22), 6371–6374. https://www.ncbi.nlm.nih.gov/pubmed/1423283.

Li, H., Fan, X., & Houghton, J. M. (2007). Tumor microenvironment: The role of the tumor stroma in cancer. *Journal of Cellular Biochemistry, 101*(4), 805–815. https://doi.org/10.1002/jcb.21159.

Lu, H., Forbes, R. A., & Verma, A. (2002). Hypoxia-inducible factor 1 activation by aerobic glycolysis implicates the Warburg effect in carcinogenesis. *Journal of Biological Chemistry, 277*(26), 23111–23115. https://doi.org/10.1074/jbc.M202487200.

Mantovani, A., Allavena, P., Sica, A., & Balkwill, F. (2008). Cancer-related inflammation. *Nature, 454*(7203), 436–444. https://doi.org/10.1038/nature07205.

Nelson, C. M., & Bissell, M. J. (2006). Of extracellular matrix, scaffolds, and signaling: Tissue architecture regulates development, homeostasis, and cancer. *Annual Review of Cell and Developmental Biology, 22*, 287–309. https://doi.org/10.1146/annurev.cellbio.22.010305.104315.

Rak, J., Yu, J. L., Klement, G., & Kerbel, R. S. (2000). Oncogenes and angiogenesis: Signaling three-dimensional tumor growth. *Journal of Investigative Dermatology - Symposium Proceedings, 5*(1), 24–33. https://doi.org/10.1046/j.1087-0024.2000.00012.x.

Semenza, G. L. (2003). Targeting HIF-1 for cancer therapy. *Nature Reviews Cancer, 3*(10), 721–732. https://doi.org/10.1038/nrc1187.

Semenza, G. L. (2012). Hypoxia-inducible factors in physiology and medicine. *Cell, 148*(3), 399–408. https://doi.org/10.1016/j.cell.2012.01.021.

Semenza, G. L. (2014). Oxygen sensing, hypoxia-inducible factors, and disease pathophysiology. *Annual Review of Pathology: Mechanisms of Disease, 9*, 47–71. https://doi.org/10.1146/annurev-pathol-012513-104720.

Siegel, R. L., Miller, K. D., & Jemal, A. (2016). Cancer statistics, 2016. *CA: A Cancer Journal for Clinicians, 66*(1), 7–30. https://doi.org/10.3322/caac.21332.

Thomlinson, R. H., & Gray, L. H. (1955). The histological structure of some human lung cancers and the possible implications for radiotherapy. *British Journal of Cancer, 9*(4), 539–549. https://doi.org/10.1038/bjc.1955.55.

Vogelstein, B., & Kinzler, K. W. (1993). The multistep nature of cancer. *Trends in Genetics, 9*(4), 138–141. https://doi.org/10.1016/0168-9525(93)90209-Z.

Weinberg, R. A. (1989). Oncogenes, antioncogenes, and the molecular bases of multistep carcinogenesis. *Cancer Research, 49*(14), 3713–3721.

Yang, J., Mani, S. A., Donaher, J. L., Ramaswamy, S., Itzykson, R. A., Come, C., et al. (2004). Twist, a master regulator of morphogenesis, plays an essential role in tumor metastasis. *Cell, 117*(7), 927–939. https://doi.org/10.1016/j.cell.2004.06.006.

CHAPTER 2

Bone marrow niches

2.1 Introduction

The term niche may be referred to the tumor microenvironment. In this context, Hambardzumyan and Bergers (2015) define niches as "communication centers in which tumor and host cell populations dynamically interact via direct cell contact or paracrine signaling cues." Bone marrow microenvironment plays an important role in the regulation of tumor growth through multiple mechanisms. It includes hematopoietic stem cells (HSCs) and nonhematopoietic cells. HSCs give rise to all the blood cell types of the myeloid and lymphoid lineages (Note 1) (Krause, 2002). The nonhematopoietic cells include endothelial cells, pericytes, fibroblasts, osteoblasts, osteoclasts, macrophages, mast cells, and mesenchymal stem cells (MSCs) (Kopp, Avecilla, Hooper, & Rafii, 2005). These cells constitute specialized niches, close to the endosteum, named "osteoblast or endosteal niche," or to the bone marrow vasculature, named "vascular niche" (Wilson & Trumpp, 2006).

2.2 Vascular niche

The blood vessels of the bone marrow not only constitute the wall that separates the hematopoietic compartment from the peripheral circulation but also are able to regulate hemopoiesis as well as stem cell mobilization and homing. Sinusoidal vessels are interconnected by intersinusoidal capillaries and collectively drain into the central sinus. In addition to providing a niche for the self-renewal, expansion, and maintenance of HSCs, sinusoidal endothelial cells play a role both in providing a differentiation platform for hematopoietic cells and a conduit for mobilization and homing hematopoietic cells into and out of the bone marrow (Fig. 2.1). The vascular niche compared with osteoblastic niche provides a more nutrient-rich microenvironment marked by higher concentrations of oxygen and growth factors, leading HSCs to be proliferative and mature blood cells, ultimately releasing into the peripheral circulation. Different

Tumor Microenvironment Regulation of Tumor Expansion
ISBN 978-0-12-822803-6
https://doi.org/10.1016/B978-0-12-822803-6.00004-1

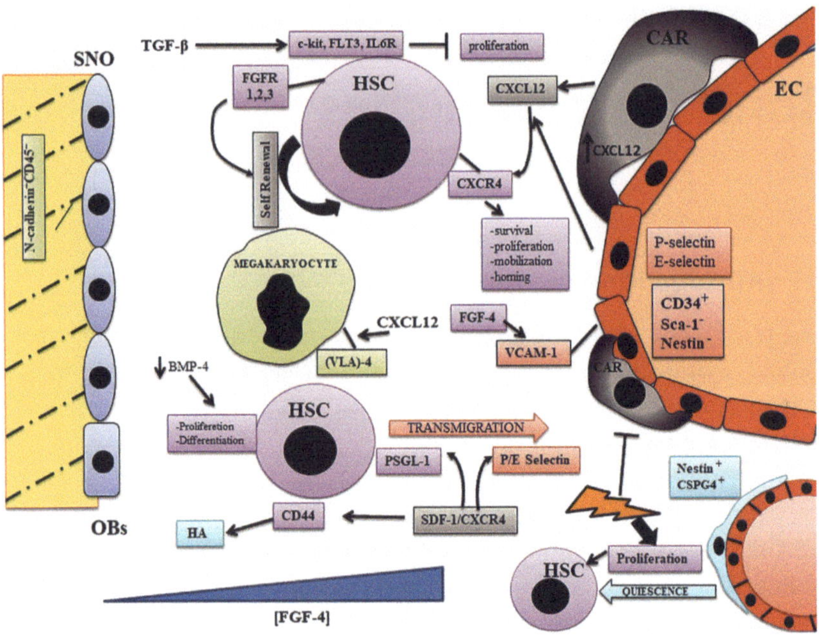

Figure 2.1 Vascular niche. In the vascular niche hematopoiesis occurs in the extravascular spaces between the sinuses. The medullary vascular sinuses are lined with endothelial cells and surrounded by adventitial cells, called CXCL12-abundant reticular (CAR) cells. The closeness between sinusoidal endothelial cells and HSCs is very important for their maturation and so for the hematopoietic process. An arteriolar niche has been found where quiescent HSCs are associated with a cell population different from CAR, named periarteriolar nestin cells. These cells express chondroitin sulfate proteoglycan-4 (CSPG4), show chemoresistance after genotoxic injury, and activate HSC proliferation. *BMP*, bone morphogenetic protein; *CARs*, CXCL12-abundant reticular cells; *CSPG4*, chondroitin sulfate proteoglycan-4; *FGF-4*, fibroblast growth factor-4; *FGFRs*, fibroblast growth factor receptors; *FLT3*, FMS like tyrosine kinase 3; *HA*, hyaluronic acid; *IL6R*, interleukin 6 receptor; *P/E selectin*, P/E selectin type P and E; *PSGL-1*, glycoprotein ligand-1; *TGF-β*, transforming growth factor-β; *VCAM-1*, vascular cell adhesion molecule 1; *VLA-4*, very late antigen. *Blue triangle*: FGF-4 gradient; lightning sign: genotoxic stimulation. *(Reproduced from Tamma and Ribatti (2017).)*

immature and mature blood cell precursors (Kusumbe et al., 2016) are in direct contact with the sinusoids $CD34^+$, nestin and Sca-1^- while endosteal endothelial cells express all these three markers (Kopp, Hooper, Avecilla, & Rafii, 2009). The stromal cells, including MSCs, form a reticular network that supports HSC formation (Benayahu, Akavia, & Shur, 2007).

Fibroblast growth factor (FGF) is an important factor in mediating the cross talk between the vascular and the endosteal niche and forms a gradient between the two bone marrow niches important in the recruitment of HSCs and their progenitors to the vascular niche expressing FGF receptors. Transforming growth factor beta (TGF-β) is a negative reversible regulator of HSCs proliferation in vitro (Batard et al., 2000). and it is thought to be a key regulator of HSC quiescence in vivo (Scandura, Boccuni, Massagué, & Nimer, 2004). High concentrations of bone morphogenetic proteins-2 and -7 (BMP-2 and BMP-7) inhibit HSC proliferation, whereas BMP-4 at high concentrations has an effect on stem cell survival, while a lower concentration stimulates stem cell proliferation and differentiation (Bhatia et al., 1999). Both the endothelial and the stromal cells express C-X-C motif chemokine 12 (CXCL12), and the CXCL12/C-X-C Motif Chemokine Receptor 4 (CXCR4) system is crucial both for cell survival and proliferation and for regulating the mobilization and the homing of HSCs. In response to CXCL12/CXCR4 signaling, endothelial cells express P-selectin and E-selectin (Katayama et al., 2003), and HSCs express glycoprotein ligand-1 (PSGL-1) which interact with both the endothelial selectins regulating the transmigration of HSCs (Katayama et al., 2003). Moreover, HSCs and progenitor cells express the smallest isoform of CD44 a cell-surface glycoprotein that is involved in cell migration for a variety of normal and malignant cells (Naor, Nedvetzki, Golan, Melnik, & Faitelson, 2002). Neutralizing antibodies anti-CD44 impaired homing of mouse hematopoietic progenitors to the bone marrow and spleen, whereas in CD44 knock-out mice a reduced myeloid progenitor cell migration from the bone marrow to blood circulation occurs (Oostendorp, Ghaffari, & Eaves, 2000; Schmits et al., 1997).

2.3 Endosteal niche

Bone tissue is constantly undergoing a process of remodeling via a tight coupling between bone formation from mesenchymal stem cell derived osteoblasts and bone resorption by osteoclasts which are hematopoietic in origin. The endosteal niche is the preferential site of HSC homing. It is localized in the internal bone shell surface (Lord, Testa, & Hendry, 1975), in particular close to the endocortical and trabecular surfaces (Askmyr, Sims, Martin, & Purton, 2009, Fig. 2.2). Endosteum is a histological structure which interposes between bone and bone marrow that include

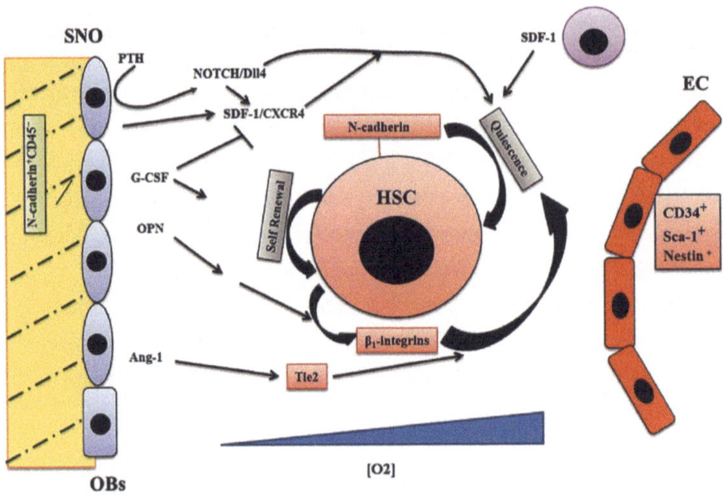

Figure 2.2 Endosteal niche. The endosteal niche is a complex structure inside which all the components, such as stem cells, progenitor cells, stromal cells, growth factors, and extracellular matrix (ECM) molecules participate in the regulation of hematopoiesis. *Ang-1*, angiopoietin-1; *CXCR4*, stromal cell derived factor-1/C-X-C motif chemokine 12; *Dll4*, δ like ligand 4; *ECs*, endothelial cells; *G-CSF*, granulocyte colony-stimulating factor; *HSCs*, hematopoietic stem cells; *NOTCH*, Notch/translocation–associated Notch homologue; *OBs*, osteoblasts; *OPN*, osteopontin; *PTH*, parathyroid hormone; *SNO*, Spindle-shaped N-cadherin$^+$CD45$^-$ osteoblastic cells; *Tie2*, angiopoietin receptor-2. *Blue triangle*: oxygen gradient. *(Reproduced from Tamma and Ribatti (2017).)*

bone-forming osteoblasts and bone-resorbing osteoclasts, and other cells including fibroblasts, macrophages, endothelial cells, adipocyte are localized near the endosteum.

There are many evidences that osteoblasts influence the HSC maturation. Bone-lining spindle-shaped osteoblasts, which can be identified by their CD45$^-$ CD31$^-$ Ter119$^-$ Sca-1$^-$ CD51$^+$ phenotype, were the first stromal cells suggested to be a crucial component of the HSC niche. Depletion of osteoblasts leads to mobilization of HSCs to the spleen (Calvi et al., 2003). Two types of osteoblastic cells are present in the endosteum, the actively bone producing osteoblasts and the quiescent bone lining cells, a type of early osteoblast termed spindle-shaped N-cadherin$^+$CD45$^-$ osteoblastic cells (SNO) (Hauge, Qvesel, Eriksen, Mosekilde, & Melsen, 2001). HSCs which adhere to osteoblasts on the bone surface express angiopoietin (Ang) receptor-2 (Tie2) while the lining cell, in adult bone marrow, produce Ang-1. The activation of Tie-2 by Ang-1 in vitro induces the adhesion of the two types of cells and consequently the immature HSCs phenotype. Moreover, Ang-1 promotes quiescence and enhances survival of HSCs and protects them from various stresses (Arai, Hirao, & Suda, 2005; Suda, Arai, & Shimmura, 2005). Osteoblasts may provide niches that support colonizing tumor cell survival while retaining them in a dormant state (Lawson et al., 2015).

The endosteal niche receives a relatively poor blood supply, maintaining HSCs in this hypoxic niche (Parmar, Mauch, Vergilio, Sackstein, & Down, 2007). The oxygen gradient guides the localization of HSCs in the endosteal region. HSCs show a selective redistribution within the endosteal region, whereas lineage-committed and mature cells are selectively redistributed toward the central marrow region so that hematopoietic differentiation seems to proceed radially toward the longitudinal axis of the marrow (Nilsson et al., 2005; Parmar et al., 2007). This could be related to the oxygen gradient into the bone marrow cavity where HSCs are predominantly located at the lowest side of the gradient because the regionally defined hypoxia plays a fundamental role in regulating stem cell function. Stem cell factor (SCF) stimulates proliferation and differentiation of HSCs in vitro (Khodadi, Shahrabi, Shahjahani, Azandeh, & Saki, 2016), and SCF receptor, c-kit, is highly expressed on quiescent HSCs in adult bone marrow (Ikuta & Weissman, 1992). Osteoblasts negatively regulates stem cell pool size (Lucas & Frenette, 2008) by the expression of osteopontin (OPN) on their membrane (Nilsson et al., 2005; Nilsson, Johnston, & Coverdale, 2001).

2.4 Bone marrow niche and vessel formation

Bone marrow in addition to its crucial role in the hematopoietic process is also involved in the process of vasculogenesis. In the embryo, the first vessels arise by de novo formation of blood vessels from scattered precursor cells, angioblasts, or vascular precursor cells, that shape blood islands which later fuse to create a primitive plexus of vessels. This process is referred to as vasculogenesis. Vasculogenesis leads to the formation of the first major intraembryonic blood vessels, such as the dorsal aorta and the cardinal veins, and to the formation of the primary vascular plexus in the yolk sac. With the onset of embryonic circulation, these primary vessels have to be remodeled into arteries and veins in order to develop a functional vascular loop. Remodeling of the primary vascular plexus into a more mature vascular system is thought to occur by angiogenesis.

Bone marrow derived endothelial cells (BMDECs) are involved in promoting indirectly (Yang et al., 2004) the vascular growth through the expression of angiogenic factors in the site where the neovascularization occurs. CXCR4 is highly expressed by BMDECs and is involved in their mobilization and homing (Burger & Kipps, 2006). CXCL12 expression is also regulated by vascular endothelial growth factor (VEGF) directly. VEGF stimulates the expression of CXCL12 in perivascular cells and the latter attracts CXCR4^{+} circulating cells. Blocking VEGF receptor-1 (VEGFR-1) reduces the number of recruited perivascular cells in tumors (Hattori et al., 2001), suggesting that the effect of VEGF could involve VEGFR-1. Bone marrow is a source for mobilized progenitor/stem cells not only involved in vasculogenesis but also responsible for the new vessel formation through angiogenesis. New vessels formed through angiogenesis during endochondral ossification are a source of perivascular osteoprogenitor cells and osteoclasts important for modeling and remodeling processes that ensure a correct skeletal development and growth (Brandi & Collin-Osdoby, 2006). Bone cells secrete proangiogenic factors, as VEGF that interact with VEGFR expressing cells, including the endothelial cells, chondrocytes, osteoblasts, and osteoclasts. In the same way endothelial cells release factors that regulate chondrocytes and cells of the osteoblast lineage (Kusumbe, Ramasamy, & Adams, 2014).

During angiogenesis that occurs in endochondral ossification, endothelial cells of the advancing capillaries directly and indirectly influence the matrix resorption by producing proteases and regulatory molecules and by recruiting osteoclast precursors from the circulation. Moreover, endothelial cells release VEGF in response to the secretion of hypoxia inducible factor 1 alpha (HIF1-α) by hypoxic chondrocytes (Manalo et al., 2005) and produce BMP-2 and BMP-4, stimulators of osteoblast differentiation. Osteoclasts are also involved in angiogenesis process. VEGF constitutes a chemoattractant factor for osteoclast precursors (Henriksen, Karsdal, Delaissé, & Engsig, 2003), and an autocrine/paracrine action of VEGF in osteoclasts has been found. During bone resorption the TGFβ1 released from the bone matrix induces VEGF expression in bone resorbing compartment that consequently stimulates endothelial activity and supports the angiogenesis. Also osteocytes contribute to angiogenesis. It is thought that during bone damage the osteocyte that undergoes apoptosis expresses VEGF (Cheung, Tonelli-Zasarsky, Simmons, & You, 2010). Osteoclasts secrete TGF-β that stimulates dormant cell reactivation and tumor growth (Croucher, McDonald, & Martin, 2016). Inhibition of bone resorption by osteoprotegerin or bisphosphonates reduces tumor volume supporting osteoclast reactivation of tumor growth in bone (Corey et al., 2003, 2005).

2.5 Note

1. Human HSCs firstly appear in the earliest embryo, then move to fetal liver and spleen, and ultimately migrate to bone marrow. Self-renewal and differentiation of HSCs are tightly regulated by both cell intrinsic and extrinsic factors and a lifelong homeostasis of functional HSCs is achieved. HSCs preferentially localize in endosteal zones, where they most closely interact with bone marrow microvessels. These special hematopoietic microenvironments are termed as "Niche," which was first proposed by Schofield in 1978, who introduced this term in hematology to designate "the cellular environment which retains the stem cell." Putative HSCs have been identified both at the endosteum, the region located in immediate proximity to bone lining osteoblasts, as well as close to sinusoidal endothelium in more central regions of the bone marrow. While some sinusoids are also found in close proximity to the endosteum, the majority is located at a greater distance to the endosteal surface, suggesting the existence of two different HSC niches in the bone marrow, an endosteal niche and a perivascular niche.

References

Arai, F., Hirao, A., & Suda, T. (2005). Regulation of hematopoietic stem cells by the niche. *Trends in Cardiovascular Medicine, 15*(2), 75–79. https://doi.org/10.1016/j.tcm.2005.03.002.

Askmyr, M., Sims, N. A., Martin, T. J., & Purton, L. E. (2009). What is the true nature of the osteoblastic hematopoietic stem cell niche? *Trends in Endocrinology and Metabolism, 20*(6), 303–309. https://doi.org/10.1016/j.tem.2009.03.004.

Batard, P., Monier, M. N., Fortunel, N., Ducos, K., Sansilvestri-Morel, P., Phan, T. H., et al. (2000). TGF-β1 maintains hematopoietic immaturity by a reversible negative control of cell cycle and induces CD34 antigen up-modulation. *Journal of Cell Science, 113*(3), 383–390.

Benayahu, D., Akavia, U. D., & Shur, I. (2007). Differentiation of bone marrow stroma-derived mesenchymal cells. *Current Medicinal Chemistry, 14*(2), 173–179. https://doi.org/10.2174/092986707779313363.

Bhatia, M., Bonnet, D., Wu, D., Murdoch, B., Wrana, J., Gallacher, L., et al. (1999). Bone morphogenetic proteins regulate the developmental program of human hematopoietic stem cells. *Journal of Experimental Medicine, 189*(7), 1139–1147. https://doi.org/10.1084/jem.189.7.1139.

Brandi, M. L., & Collin-Osdoby, P. (2006). Vascular biology and the skeleton. *Journal of Bone and Mineral Research, 21*(2), 183–192. https://doi.org/10.1359/JBMR.050917.

Burger, J. A., & Kipps, T. J. (2006). CXCR4: A key receptor in the crosstalk between tumor cells and their microenvironment. *Blood, 107*(5), 1761–1767. https://doi.org/10.1182/blood-2005-08-3182.

Calvi, L. M., Adams, G. B., Weibrecht, K. W., Weber, J. M., Olson, D. P., Knight, M. C., et al. (2003). Osteoblastic cells regulate the haematopoietic stem cell niche. *Nature, 425*(6960), 841–846. https://doi.org/10.1038/nature02040.

Cheung, W.-Y., Liu, C., Tonelli-Zasarsky, R. M. L., Simmons, C. A., & You, L. (2010). Osteocyte apoptosis is mechanically regulated and induces angiogenesis in vitro. *Journal of Orthopaedic Research, 29*(4), 523–530. https://doi.org/10.1002/jor.21283.

Corey, E., Brown, L. G., Kiefer, J. A., Quinn, J. E., Pitts, T. E. M., Blair, J. M., et al. (2005). Osteoprotegerin in prostate cancer bone metastasis. *Cancer Research, 65*(5), 1710–1718. https://doi.org/10.1158/0008-5472.CAN-04-2033.

Corey, E., Brown, L. G., Quinn, J. E., Poot, M., Roudier, M. P., Higano, C. S., et al. (2003). Zoledronic acid exhibits inhibitory effects on osteoblastic and osteolytic metastases of prostate cancer. *Clinical Cancer Research, 9*(1 I), 295–306.

Croucher, P. I., McDonald, M. M., & Martin, T. J. (2016). Bone metastasis: The importance of the neighbourhood. *Nature Reviews Cancer, 16*(6), 373–386. https://doi.org/10.1038/nrc.2016.44.

Hambardzumyan, D., & Bergers, G. (2015). Glioblastoma: Defining tumor niches. *Trends in Cancer, 1*(4), 252–265. https://doi.org/10.1016/j.trecan.2015.10.009.

Hattori, K., Dias, S., Heissig, B., Hackett, N. R., Lyden, D., Tateno, M., et al. (2001). Vascular endothelial growth factor and angiopoietin-1 stimulate postnatal hematopoiesis by recruitment of vasculogenic and hematopoietic stem cells. *Journal of Experimental Medicine, 193*(9), 1005–1014. https://doi.org/10.1084/jem.193.9.1005.

Hauge, E. M., Qvesel, D., Eriksen, E. F., Mosekilde, L., & Melsen, F. (2001). Cancellous bone remodeling occurs in specialized compartments lined by cells expressing osteoblastic markers. *Journal of Bone and Mineral Research, 16*(9), 1575–1582. https://doi.org/10.1359/jbmr.2001.16.9.1575.

Henriksen, K., Karsdal, M., Delaissé, J. M., & Engsig, M. T. (2003). RANKL and vascular endothelial growth factor (VEGF) induce osteoclast chemotaxis through an ERK1/2-dependent mechanism. *Journal of Biological Chemistry, 278*(49), 48745–48753. https://doi.org/10.1074/jbc.M309193200.

Ikuta, K., & Weissman, I. L. (1992). Evidence that hematopoietic stem cells express mouse c-kit but do not depend on steel factor for their generation. *Proceedings of the National Academy of Sciences of the United States of America, 89*(4), 1502–1506. https://doi.org/10.1073/pnas.89.4.1502.

Katayama, Y., Hidalgo, A., Furie, B. C., Vestweber, D., Furie, B., & Frenette, P. S. (2003). PSGL-1 participates in E-selectin-mediated progenitor homing to bone marrow: Evidence for cooperation between E-selectin ligands and α4 integrin. *Blood, 102*(6), 2060–2067. https://doi.org/10.1182/blood-2003-04-1212.

Khodadi, E., Shahrabi, S., Shahjahani, M., Azandeh, S., & Saki, N. (2016). Role of stem cell factor in the placental niche. *Cell and Tissue Research, 366*(3), 523–531. https://doi.org/10.1007/s00441-016-2429-3.

Kopp, H. G., Avecilla, S. T., Hooper, A. T., & Rafii, S. (2005). The bone marrow vascular niche: Home of HSC differentiation and mobilization. *Physiology, 20*(5), 349–356. https://doi.org/10.1152/physiol.00025.2005.

Kopp, H. G., Hooper, A. T., Avecilla, S. T., & Rafii, S. (2009). Functional heterogeneity of the bone marrow vascular niche. *Annals of the New York Academy of Sciences, 1176*, 47–54. https://doi.org/10.1111/j.1749-6632.2009.04964.x.

Krause, D. S. (2002). Regulation of hematopoietic stem cell fate. *Oncogene, 21*(21), 3262–3269. https://doi.org/10.1038/sj.onc.1205316.

Kusumbe, A. P., Ramasamy, S. K., & Adams, R. H. (2014). Coupling of angiogenesis and osteogenesis by a specific vessel subtype in bone. *Nature, 507*(7492), 323–328. https://doi.org/10.1038/nature13145.

Kusumbe, A. P., Ramasamy, S. K., Itkin, T., Mäe, M. A., Langen, U. H., Betsholtz, C., et al. (2016). Age-dependent modulation of vascular niches for haematopoietic stem cells. *Nature, 532*(7599), 380–384. https://doi.org/10.1038/nature17638.

Lawson, M. A., McDonald, M. M., Kovacic, N., Hua Khoo, W., Terry, R. L., Down, J., et al. (2015). Osteoclasts control reactivation of dormant myeloma cells by remodelling the endosteal niche. *Nature Communications, 6*, 8983. https://doi.org/10.1038/ncomms9983.

Lord, B. I., Testa, N. G., & Hendry, J. H. (1975). The relative spatial distributions of CFU(s) and CFU(c) in the normal mouse femur. *Blood, 46*(1), 65–72.

Lucas, D., & Frenette, P. S. (2008). Osteoblasts: Yes, they can. *Blood, 112*(3), 455. https://doi.org/10.1182/blood-2008-05-158758.

Manalo, D. J., Rowan, A., Lavoie, T., Natarajan, L., Kelly, B. D., Ye, S. Q., et al. (2005). Transcriptional regulation of vascular endothelial cell responses to hypoxia by HIF-1. *Blood, 105*(2), 659–669. https://doi.org/10.1182/blood-2004-07-2958.

Naor, D., Nedvetzki, S., Golan, I., Melnik, L., & Faitelson, Y. (2002). CD44 in cancer. *Critical Reviews in Clinical Laboratory Sciences, 39*(6), 527–579. https://doi.org/10.1080/10408360290795574.

Nilsson, S. K., Johnston, H. M., & Coverdale, J. A. (2001). Spatial localization of transplanted hemopoietic stem cells: Inferences for the localization of stem cell niches. *Blood, 97*(8), 2293–2299. https://doi.org/10.1182/blood.V97.8.2293.

Nilsson, S. K., Johnston, H. M., Whitty, G. A., Williams, B., Webb, R. J., Denhardt, D. T., et al. (2005). Osteopontin, a key component of the hematopoietic stem cell niche and regulator of primitive hematopoietic progenitor cells. *Blood, 106*(4), 1232–1239. https://doi.org/10.1182/blood-2004-11-4422.

Oostendorp, R. A. J., Ghaffari, S., & Eaves, C. J. (2000). Kinetics of in vivo homing and recruitment into cycle of hematopoietic cells are organ-specific but CD44-independent. *Bone Marrow Transplantation, 26*(5), 559–566. https://doi.org/10.1038/sj.bmt.1702536.

Parmar, K., Mauch, P., Vergilio, J.-A., Sackstein, R., & Down, J. D. (2007). Distribution of hematopoietic stem cells in the bone marrow according to regional hypoxia. *Proceedings of the National Academy of Sciences of the United States of America, 104*(13), 5431–5436. https://doi.org/10.1073/pnas.0701152104.

Scandura, J. M., Boccuni, P., Massagué, J., & Nimer, S. D. (2004). Transforming growth factor β-induced cell cycle arrest of human hematopoietic cells requires p57KIP2 up-regulation. *Proceedings of the National Academy of Sciences of the United States of America, 101*(42), 15231–15236. https://doi.org/10.1073/pnas.0406771101.

Schmits, R., Filmus, J., Gerwin, N., Senaldi, G., Kiefer, F., Kundig, T., et al. (1997). CD44 regulates hematopoietic progenitor distribution, granuloma formation, and tumorigenicity. *Blood, 90*(6), 2217–2233. https://doi.org/10.1182/blood.v90.6.2217.2217_2217_2233.

Schofield, R. (1978). The relationship between the spleen colony forming cell and the haemopoietic stem cell. *Blood Cells, 4*, 7–25.

Suda, T., Arai, F., & Shimmura, S. (2005). Regulation of stem cells in the niche. *Cornea, 24*(8), S12–S17.

Tamma, R., & Ribatti, D. (2017). Bone niches, hematopoietic stem cells and vessel formation. *International Journal of Molecular Sciences, 18*, 151. https://doi.org/10.3390/ijms18010151.

Wilson, A., & Trumpp, A. (2006). Bone-marrow haematopoietic-stem-cell niches. *Nature Reviews Immunology, 6*(2), 93–106. https://doi.org/10.1038/nri1779.

Yang, L., DeBusk, L. M., Fukuda, K., Fingleton, B., Green-Jarvis, B., Shyr, Y., et al. (2004). Expansion of myeloid immune suppressor Gr+CD11b+ cells in tumor-bearing host directly promotes tumor angiogenesis. *Cancer Cell, 6*(4), 409–421. https://doi.org/10.1016/j.ccr.2004.08.031.

CHAPTER 3

Metastatic cascade

3.1 Introduction

In 1829, Joseph Récaimer was the first to use the term metastasis, referred to the spread of tumor cells along external surfaces of vascular channels rather than within them. In almost solid tumors, the biggest cause of mortality are metastases (Steegs, 2006). Approximately 30% of newly diagnosed solid tumors present with metastases. Blood cancers, sometimes called liquid tumors, are derived from blood-forming cells that have the capacity to enter the bloodstream and travel to distant sites. Patients die of metastatic disease, and prevention of metastasis and treatment of micrometastatic disease is extremely important to improve cure rates. Premetastatic tumor cells migrate from the primary tumor and intravasate into the circulation. Tumor cells that resisted the various types of pressure in the circulation, extravasate, colonize organs, and then generate and shape a metastatic microenvironment specific to that organ (Fidler, 2003). Given the dynamic, multistep nature of metastasis, and the presence of intratumor heterogeneity, certain cancer cell subpopulations may potentially perform some steps of metastasis more efficiently than others.

The microenvironment of the primary tumor as well as the metastatic microenvironment of the secondary organs have the capacity to regulate tumor-intrinsic features. The organs mostly involved in metastases are lung, liver, brain, and bone. The lungs serve as the first filter for tumor cells spreading through blood circulation in malignancies whose venous drainage flows directly into the lungs. Testis, melanoma, osteosarcoma, and head and neck tumors have the highest incidence of pulmonary metastases. The liver is one of the most common sites for metastatic disease, accounting for 25% of all metastases to solid organs. The most frequent metastases to the brain occurs in patients with lung, breast, melanoma, renal, and colorectal tumors. Bone metastases are most commonly seen in prostate, breast, and lung cancer, which are leading malignancies in female and/or male having the highest incidence and mortality rates. During the development of bone metastases from prostate cancer, the interactions among tumor cells, bone

Tumor Microenvironment Regulation of Tumor Expansion
ISBN 978-0-12-822803-6
https://doi.org/10.1016/B978-0-12-822803-6.00001-6

Table 3.1 Different stages of the metastatic process.

Invasion/migration at/near the primary tumor
Intravasation into the local blood and lymphatic vessels
Survival ant transit of cancer cells in the circulation/lymphatics
Arrest and extravasation at secondary sites
Colonization of secondary sites

cells, and bone matrix constitute a "vicious cycle" of osteoblast/osteoclast-mediated bone metastasis. Tumor cells influence the stroma, and vice versa, jointly driving cancer progression. The invasion and metastatic process are the final result of a complex series of interactions between cells and the microenvironment that surround tumor, based on the ability to modify tissue architecture through degradation and remodeling of the extracellular matrix (ECM) (Table 3.1) (Quail & Joyce, 2013).

3.2 Mechanisms of metastasis

Mechanisms of metastasis involve a complex array of genetic and epigenetic changes specific both for different types of tumors and for different site of metastasis. Hypoxia can increase the metastatic potential of a tumor through genetic and epigenetic alterations acquired in order to adapt to the hypoxic microenvironment, and also through the expression of molecules that disrupt or modify the ability of the tumor cells to interact with the ECM facilitating the metastatic process. The efficiency of metastasis is low; it is estimated to be less than 0.01 per cent (Fidler, 1970). Only a small proportion of tumor cells released into the circulation form macroscopic metastasis. Intravital microscopy studies have shown that for some tumor cells, extravasation can be efficient, allowing for >80 per cent cells that enter the circulation to move from the vasculature into the interstitial space. Only a small proportion of these cells form micrometastases, and only a proportion of these became vascularized and grew into macroscopic metastases (Chambers, Groom, & MacDonald, 2002). Consistent with this notion, among 1438 breast carcinoma patients who harbored hundreds to thousands of micrometastases in their bone marrow at the time of initial diagnosis, only 50 per cent developed clinically detectable metastases within 10 years (Braun et al., 2005). Breast cancer develop metastasis years to decades following removal of the primary tumor mass (Karrison, Ferguson, & Meier, 1999).

A number of metastasis suppressor genes have been identified that can suppress growth at a secondary site without suppressing primary tumor growth (Yoshida, Sokoloff, Welch, & Rinker-Schaeffer, 2000). Apoptosis can be an important determinant for metastatic inefficiency, as a significant higher number of apoptotic cells was demonstrated among cells arrested in the lung following intravenous injection of a low-metastatic versus high-metastatic cell line (Wong et al., 2001). In 1889, Stephen Paget (Fig. 3.1) proposed in a study conducted on autopsy records of 735 women who had died of breast cancer, that the process of metastasis did not occur by chance but, rather, that certain favored tumor cells, with metastatic activity (the "seed") had a special affinity for the growth-enhancing milieu within specific organs (the "soil"), i.e., organs providing a growth advantage to the seeds. He concluded that metastases developed only when the seed and soil were compatible (Paget, 1989). In other words, Paget suggested that the site of metastasis depended on the affinity of the tumor to the microenvironment, and this theory has been widely accepted as a basic principle in the field of cancer metastasis. Paget's observations contradicted the theory of Rudolf Virchow (Fig. 3.2) that metastasis could be explained simply by the arrest of tumor cell emboli in the vasculature (Virchow, 1858). Consistent with the seed-and-soil hypothesis, evidence from a number of laboratories has documented that specific organ microenvironments are indeed

Figure 3.1 A portrait of Stephen Paget. *(From Wikipedia.)*

Figure 3.2 A portrait of Rudolf Virchow. *(From Wikipedia.)*

intrinsically more or less hospitable for the proliferation and survival of certain types of disseminated tumor cells. For example, melanoma cells readily metastasized to subcutaneous grafts of lung tissue but failed to metastasize to identically placed and comparably vascularized subcutaneous grafts of renal tissue.

In 1928, James Ewing (Fig. 3.3) proposed that metastatic dissemination of cancer was dependent on the anatomical distribution of the vascular

Figure 3.3 A port trait of James Ewing. *(From Wikipedia.)*

Figure 3.4 A portrait of Isaiah Fidler.

system (Ewing, 1928). This controversy was resolved by Isaiah Fidler (Fig. 3.4) and collaborators, who studied experimental metastasis in syngeneic mice to show that subsequent metastatic growth at a distant organ site was site-specific, consistent with Paget's original hypothesis (Hart, 1979). Fidler demonstrated that certain kinds of cancer metastasize to specific organs regardless of blood flow or vascular anatomy (Fidler, 2003). It has long been accepted that most malignant tumors show an organ-specific pattern of metastasis. For example, colon carcinomas metastasize usually to liver and lung but rarely to bone, skin, or brain and almost never to kidneys, intestine, or muscle. In contrast, other tumor entities, such as breast carcinomas, frequently form metastases in most of these organs.

Another concept of metastasis formation is focusing on the systemic signaling that occurs between primary tumors and metastases. McAllister et al. (2008) demonstrated that a primary tumor ("instigator") can stimulate the outgrowth of distant, indolent tumor cells ("responders") and named this process "systemic instigation." In their study, nude mice with green flurescent protein (GFP)-positive bone marrow served as hosts for xenograft tumors of breast cancer, and GFP-labeled bone marrow derived cells were only recruited into responding tumors in the presence of an instigating tumor. Osteopontin (OPN) released in the circulation was shown to be necessary for the recruitment and the subsequent outgrowth of indolent tumors (McAllister et al., 2008). The specific formation of secondary tumors at distant sites must be successfully completed by metastasizing tumor cells (Chambers et al., 2002).

Various explanations have been proposed for the site selectivity of blood-borne metastases, including tumor cell surface characteristics (Turner, 1982), response to organ-derived chemotactic factors (Hujanen & Terranova, 1985), adhesion between tumor cells and the target organ components (Nicolson, 1988), and response to specific host tissue growth factors (Nicolson & Dulski, 1986). The relative importance of preexisting tumor subpopulations with specific metastatic properties and the organ environment characteristics in determining metastatic homing have been debated (Talmadge & Fidler, 1982). Murine melanomas grown in a cranial window have higher vessel density and branching as compared to those in the same tumors grown in a dorsal skin chamber (Kashiwagi et al., 2005). Human renal cell carcinoma xenografts grown orthotopically in the kidneys of immunodeficient mice were highly vascularized and metastatic than those from the same tumor grown subcutaneously (Singh et al., 1994). Other anatomic structures are used as a scaffolding for cancer cell migration, including nerves (Chen et al., 2019), the peritoneum, pleura, myofibers, adipocytes, bone cavities, brain ventricles, choroid plexuses, and the glia limitans in the brain (Friedl & Alexander, 2011). Furthermore, remodeling of the ECM by cancer-associated fibroblasts generates tracks for cancer cells (Erdogan et al., 2017). An alternative explanation for the different sites of tumor growth involves interactions between the metastatic cells and the organ environment, in terms of specific binding to endothelial cells and responses to local growth factors. The mechanisms by which cancer cells can transmigrate through the endothelial lining are not well understood. Signaling cross talk between cancer cells and endothelial cells may involve upregulation of adhesion molecule expression by the endothelium as well as by the tumor cells, reorganization of the actomyosin cytoskeleton, and Src-mediated disruption of endothelial vascular endothelial (VE)-cadherin-β-catenin cell—cell adhesion. Endothelial cells in the vasculature of different organs express different cell-surface receptors and growth factors that influence the phenotype of the corresponding metastases. Greene and Harvey (1964) suggested that the organ distribution patterns of metastatic foci were dependent on the formation of sufficient adhesive bonds between arrested tumor cells and endothelial cells, and they hypothesized that these interactions were similar to lymphocyte/endothelial cells at sites of inflammation.

The development of organ-derived microvascular endothelial cell cultures has allowed more specific studies on the preferential homing of tumor cells. Auerbach, Alby, Morrissey, Tu, and Joseph (1985), Auerbach et al. (1987) found that different tumors showed differences in their adhesive propensity and preference for different endothelial cells, and in a few cases preferential adhesion was observed to the endothelial cells derived from the organ of origin and the target organ. The venous blood that circulates through most organs drains back to the lungs, explaining why the lungs are such a common site of metastasis. Blood from the colon and gastrointestinal tract drains first into the liver, which is frequently the initial site of colorectal cancer metastasis (Denève et al., 2013). Apart from metastases in the lungs and liver, which are initially seeded via the portal vein and systemic veins, respectively, and skeletal metastases seeded via the paravertebral venous plexus, hematogenous metastases are seeded via arteries. Certain cancers have the propensity for invasion of veins. For example, renal cell carcinoma often invades the branches of the renal vein and then the renal vein itself.

3.3 Lymphatic vessels

Tumors do not contain functional lymphatics, but lymphatic vessels located at the tumor margin are sufficient for the lymphatic spread of tumor cells. The primary function of lymphatics is not to carry oxygen or essential nutrients. They instead absorb extravasated protein-rich fluids, lipids, macromolecules, and immunocompetent cells from the interstitial spaces within tissues. Normal and functioning lymphatic vessels thus maintain plasma volume, prevent increases in tissue pressure, and allow leukocyte trafficking, thereby also playing a key role in the proper functioning of the immune system. In tumors, lymphatic vessels are unable to remove interstitial fluid contributing to elevated interstitial fluid pressure and edema (Jain, Tong, & Munn, 2007). Moreover, vascular endothelial growth factor-C (VEGF-C) render lymphatic vessels more permeable (Tammela et al., 2007). Blocking VEGF receptor-3 (VEGFR-3) signaling reduced collecting vessel dilatation and decreased lymph flow and nodal metastasis (Karnezis et al., 2012).

In a number of cancer types, including breast cancer, melanoma, prostate cancer, gastric cancer, and colon cancer, the occurrence of metastasis in the tumor draining lymph node (the "sentinel" lymph node) (Note 1) is regarded the first step in metastatic cancer dissemination. During metastasis, cancer cells escape the primary tumor, extravasate into lymphatic vessels, and reach draining sentinel lymph node well before they appear to colonize distant organs via the blood circulation (Brown et al., 2018). Tumor-induced changes in the sentinel lymph node include changes in the immune composition of the lymph node, in terms of changes in the density and maturation state of dendritic cells and T cells, in the cytokine composition, and the ability of the immune cells to generate a specific immune response (Cochran et al., 2006). Tumor cells can interact either directly or indirectly with the immune cells within the lymph nodes by means of tumor secreted factors or extravascular vesicles (Jung et al., 2009). Lymphangiogenesis can occur in regional lymph nodes before metastasis has occurred, creating a favorable environment for disseminated cells (Alitalo & Detmar, 2011). Peritumoral lymphatics are significantly higher than intratumoral ones (Sleeman & Thiele, 2009), that are often absent (Padera et al., 2002). Peritumoral lymphatic vessel density is a predictor of lymph node metastasis and poor prognosis (Zhang, Zhang, & Yi, 2016). There are numerous interconnections between the lymphatic and vascular systems. Cells from lymph node metastases can be carried in efferent lymph vessels to the venous circulation and colonize organs by hematogenous spread. Tumor growth stimulates lymphangiogenesis, and tumor-associated lymphatics represent a dissemination route for metastatic tumor cells to draining lymph nodes (Dieterich & Detmar, 2016).

3.4 Premetastatic niche

Before metastatic dissemination, primary tumors secrete factors that contribute to the development of a premetastatic niche (Note 2) characterized by the presence of an increased number of fibroblasts and secreted cytokines rendering the microenvironment receptive to tumor growth. The formation of a supportive premetastatic niche could represent a broadly important determinant of both metastatic propensity and tissue tropism. Bone marrow–derived VEGFR1$^+$ cells interfered with the formation of these clusters and blocked metastasis (Kaplan et al., 2005). Premetastatic niches have been found in the lymph nodes of patients with colorectal, prostate, breast, thyroid, bladder, gastric, and renal cell carcinomas (Sleeman, 2015). At secondary sites, angiogenic dormancy is characterized by avascular micrometastatic lesions that do not grow beyond 1–2 mm in diameter because of balance between proliferation and apoptosis. A dormant niche is associated with mature vessels and endothelial derived thrombospondin-1 (TSP-1), whereas metastatic outgrowth is associated with sprouting vessels and production of periostin and transforming growth factor beta (TGF-β) (Ghajar et al., 2013). Exosomes (Note 3) are released from both cancer and normal host cells that provide signals for intercellular communication within tumors and between tumors and normal organs, and promote metastatic spread in the lung, bone, liver, and brain (Hoshino et al., 2015; Lakkaraju & Rodriguez-Boulan, 2008; Van Niel, Porto-Carreiro, Simoes, & Raposo, 2006). Exosomes released by tumor cells fuse with stromal cells, including fibroblasts, Kupffer cells, and endothelial cells (Hoshino et al., 2015). The vesicles contain a variety of cytokines, chemokines, that trigger a local inflammatory response, angiogenesis, and tissue remodeling (Liu & Cao, 2016).

3.5 Notes

1. The first lymph node(s) to which cancer cells are likely to spread from a primary tumor. To identify the sentinel lymph node(s), the surgeon injects a radioactive substance, blue dye, or both near the tumor. The surgeon then uses a probe to find the sentinel lymph node(s) containing the radioactive substance or looks for the lymph node(s) stained with dye. The surgeon then removes the sentinel node(s) to check for the presence of cancer cells. In many cases, the regional lymph nodes serve as effective barriers to further dissemination of the tumor, and after the arrest within the node the cells may be destroyed by a tumor-specific immune response. In the presence of a tumor, the afferent lymphatic vessel leading into the sentinel lymph node becomes dilated, and the subcapsular sinus opens and in the collecting lymphatic vessels, tumor cells are carried passively with the flow of lymph toward the lymph node, as single cells and clusters (emboli).
2. The premetastatic niche refers to the microenvironment, which is well prepared for tumor cells to colonize in and disseminate to distant organ sites, prior to the arrival of potentially metastatic cells. Lyden firstly proposed the idea of the premetastatic niche (Kaplan et al., 2005). Removal of the primary tumor should result in the cessation of formation of the premetastatic niche.
3. Exosomes are nano-sized biovesicles released into surrounding body fluids upon fusion of multivesicular bodies and the plasma membrane. They were shown to carry cell-specific cargos of proteins, lipids, and genetic materials, and can be selectively taken up by neighboring or distant cells far from their release, reprogramming the recipient cells upon their bioactive compounds.

References

Alitalo, A., & Detmar, M. (2011). Interactions of tumor cells and lymphatic vessels in cancer progression. *Oncogene, 31*, 4499–4508.

Auerbach, R., Alby, L., Morrissey, L. W., Tu, M., & Joseph, J. (1985). Expression of organ-specific antigens on capillary endothelial cells. *Microvascular Research, 29*(3), 401–411. https://doi.org/10.1016/0026-2862(85)90028-7.

Auerbach, R., Lu, W. C., Pardon, E., Gumkowski, F., Kaminska, G., & Kaminski, M. (1987). Specificity of adhesion between murine tumor cells and capillary endothelium: An in vitro correlate of preferential metastasis in vivo. *Cancer Research, 47*(6), 1492–1496. https://www.ncbi.nlm.nih.gov/pubmed/3815350.

Braun, S., Vogl, F. D., Naume, B., Janni, W., Osborne, M. P., Coombes, R. C., et al. (2005). A pooled analysis of bone marrow micrometastasis in breast cancer. *New England Journal of Medicine, 353*(8), 793–802. https://doi.org/10.1056/NEJMoa050434.

Brown, M., Assen, F. P., Leithner, A., Abe, J., Schachner, H., Asfour, G., et al. (2018). Lymph node blood vessels provide exit routes for metastatic tumor cell dissemination in mice. *Science, 359*(6382), 1408–1411. https://doi.org/10.1126/science.aal3662.

Chambers, A. F., Groom, A. C., & MacDonald, I. C. (2002). Dissemination and growth of cancer cells in metastatic sites. *Nature Reviews Cancer, 2*(8), 563–572. https://doi.org/10.1038/nrc865.

Chen, S. H., Zhang, B. Y., Zhou, B., Zhu, C. Z., Sun, L. Q., & Feng, Y. J. (2019). Perineural invasion of cancer: A complex crosstalk between cells and molecules in the perineural niche. *American Journal of Cancer Research, 9*(1), 1–21. https://www.ncbi.nlm.nih.gov/pubmed/30755808.

Cochran, A. J., Huang, R. R., Lee, J., Itakura, E., Leong, S. P. L., & Essner, R. (2006). Tumour-induced immune modulation of sentinel lymph nodes. *Nature Reviews Immunology, 6*(9), 659–670. https://doi.org/10.1038/nri1919.

Denève, E., Riethdorf, S., Ramos, J., Nocca, D., Coffy, A., Daurès, J. P., et al. (2013). Capture of viable circulating tumor cells in the liver of colorectal cancer patients. *Clinical Chemistry, 59*(9), 1384–1392. https://doi.org/10.1373/clinchem.2013.202846.

Dieterich, L. C., & Detmar, M. (2016). Tumor lymphangiogenesis and new drug development. *Advanced Drug Delivery Reviews, 99*, 148–160. https://doi.org/10.1016/j.addr.2015.12.011.

Erdogan, B., Ao, M., White, L. M., Means, A. L., Brewer, B. M., Yang, L., et al. (2017). Cancer-associated fibroblasts promote directional cancer cell migration by aligning fibronectin. *The Journal of Cell Biology, 216*(11), 3799–3816. https://doi.org/10.1083/jcb.201704053.

Ewing, J. (1928). Neoplastic diseases. In *A treatise on tumors* (2nd ed. revised, pp. 1–1127). Philadelphia: W B Saunders and Co.

Fidler, I. J. (1970). Metastasis: Quantitative analysis of distribution and fate of tumor emboli labeled with 125l-5-lodo-2'-deoxyuridine. *Journal of the National Cancer Institute, 45*(4), 773–782. https://doi.org/10.1093/jnci/45.4.773.

Fidler, I. J. (2003). The pathogenesis of cancer metastasis: The "seed and soil" hypothesis revisited. *Nature Reviews Cancer, 3*(6), 453–458. https://doi.org/10.1038/nrc1098.

Friedl, P., & Alexander, S. (2011). Cancer invasion and the microenvironment: Plasticity and reciprocity. *Cell, 147*(5), 992–1009. https://doi.org/10.1016/j.cell.2011.11.016.

Ghajar, C. M., Peinado, H., Mori, H., Matei, I. R., Evason, K. J., Brazier, H., et al. (2013). The perivascular niche regulates breast tumour dormancy. *Nature Cell Biology, 15*(7), 807–817. https://doi.org/10.1038/ncb2767.

Greene, H. S. N., & Harvey, E. K. (1964). The relationship between the dissemination of tumor cells and the distribution of metastases. *Cancer Research, 24*(5), 799–811.

Hart, I. R. (1979). The selection and characterization of an invasive variant of the B16 melanoma. *American Journal of Pathology, 97*(3), 587–600.
Hoshino, A., Costa-Silva, B., Shen, T.-L., Rodrigues, G., Hashimoto, A., Tesic Mark, M., et al. (2015). Tumour exosome integrins determine organotropic metastasis. *Nature, 527*(7578), 329–335. https://doi.org/10.1038/nature15756.
Hujanen, E. S., & Terranova, V. P. (1985). Migration of tumor cells to organ-derived chemoattractants. *Cancer Research, 45*(8), 3517–3521.
Jain, R. K., Tong, R. T., & Munn, L. L. (2007). Effect of vascular normalization by antiangiogenic therapy on interstitial hypertension, peritumor edema, and lymphatic metastasis: Insights from a mathematical model. *Cancer Research, 67*(6), 2729–2735. https://doi.org/10.1158/0008-5472.CAN-06-4102.
Jung, T., Castellana, D., Klingbeil, P., Hernández, I. C., Vitacolonna, M., Orlicky, D. J., et al. (2009). CD44v6 dependence of premetastatic niche preparation by exosomes. *Neoplasia, 11*(10), 1093–1105. https://doi.org/10.1593/neo.09822.
Kaplan, R. N., Riba, R. D., Zacharoulis, S., Bramley, A. H., Vincent, L., Costa, C., et al. (2005). VEGFR1-positive haematopoietic bone marrow progenitors initiate the pre-metastatic niche. *Nature, 438*(7069), 820–827. https://doi.org/10.1038/nature04186.
Karnezis, T., Shayan, R., Caesar, C., Roufail, S., Harris, N. C., Ardipradja, K., et al. (2012). VEGF-D promotes tumor metastasis by regulating prostaglandins produced by the collecting lymphatic endothelium. *Cancer Cell, 21*(2), 181–195. https://doi.org/10.1016/j.ccr.2011.12.026.
Karrison, T. G., Ferguson, D. J., & Meier, P. (1999). Dormancy of mammary carcinoma after mastectomy. *Journal of the National Cancer Institute, 91*(1), 80–85. https://doi.org/10.1093/jnci/91.1.80.
Kashiwagi, S., Izumi, Y., Gohongi, T., Demou, Z. N., Xu, L., Huang, P. L., et al. (2005). NO mediates mural cell recruitment and vessel morphogenesis in murine melanomas and tissue-engineered blood vessels. *Journal of Clinical Investigation, 115*(7), 1816–1827. https://doi.org/10.1172/JCI24015.
Lakkaraju, A., & Rodriguez-Boulan, E. (2008). Itinerant exosomes: Emerging roles in cell and tissue polarity. *Trends in Cell Biology, 18*(5), 199–209. https://doi.org/10.1016/j.tcb.2008.03.002.
Liu, Y., & Cao, X. (2016). Nicolson GL 1988. Organ specificity of tumor metastasis: Role of preferential adhesion, invasion and growth of malignant cells at specific secondary sites. *Cancer and Metastasis Reviews, 30*, 143–188.
McAllister, S. S., Gifford, A. M., Greiner, A. L., Kelleher, S. P., Saelzler, M. P., Ince, T. A., et al. (2008). Systemic endocrine instigation of indolent tumor growth requires osteopontin. *Cell, 133*(6), 994 1005. https://doi.org/10.1016/j.cell.2008.04.045.
Nicolson, G. L. (1988). Organ specificity of tumor metastasis: Role of preferential adhesion, invasion and growth of malignant cells at specific secondary sites. *Cancer and Metastasis Review, 7*(2), 143–188. https://doi.org/10.1007/BF00046483.
Nicolson, G. L., & Dulski, K. M. (1986). Organ specificity of metastatic tumor colonization is related to organ-selective growth properties of malignant cells. *International Journal of Cancer, 38*(2), 289–294. https://doi.org/10.1002/ijc.2910380221.
Padera, T. P., Kadambi, A., Di Tomaso, E., Mouta Carreira, C., Brown, E. B., Boucher, Y., et al. (2002). Lymphatic metastasis in the absence of functional intratumor lymphatics. *Science, 296*(5574), 1883–1886. https://doi.org/10.1126/science.1071420.
Paget, S. (1989). The distribution of secondary growths in cancer of the breast. 1889. *Cancer and Metastasis Reviews, 8*(2), 98–101. https://www.ncbi.nlm.nih.gov/pubmed/2673568.
Quail, D. F., & Joyce, J. A. (2013). Microenvironmental regulation of tumor progression and metastasis. *Nature Medicine, 19*(11), 1423–1437. https://doi.org/10.1038/nm.3394.

Singh, R. K., Bucana, C. D., Gutman, M., Fan, D., Wilson, M. R., & Fidler, I. J. (1994). Organ site-dependent expression of basic fibroblast growth factor in human renal cell carcinoma cells. *American Journal of Pathology, 145*(2), 365–374.

Sleeman, J. P. (2015). The lymph node pre-metastatic niche. *Journal of Molecular Medicine, 93*(11), 1173–1184. https://doi.org/10.1007/s00109-015-1351-6.

Sleeman, J. P., & Thiele, W. (2009). Tumor metastasis and the lymphatic vasculature. *International Journal of Cancer, 125*(12), 2747–2756. https://doi.org/10.1002/ijc.24702.

Steegs, P. S. (2006). Tumor metastasis: mechanistic insights and clinical challenges. *Nature Medicine, 12*(8), 895–904. https://doi.org/10.1038/nm1469.

Talmadge, J. E., & Fidler, I. J. (1982). Cancer metastasis is selective or random depending on the parent tumour population. *Nature, 297*(5867), 593–594. https://doi.org/10.1038/297593a0.

Tammela, T., Saaristo, A., Holopainen, T., Lyytikkä, J., Kotronen, A., Pitkonen, M., et al. (2007). Therapeutic differentiation and maturation of lymphatic vessels after lymph node dissection and transplantation. *Nature Medicine, 13*(12), 1458–1466. https://doi.org/10.1038/nm1689.

Turner, G. A. (1982). Surface properties of the metastatic cell. *Invasion and Metastasis, 2*(4), 197–216.

Van Niel, G., Porto-Carreiro, I., Simoes, S., & Raposo, G. (2006). Exosomes: A common pathway for a specialized function. *Journal of Biochemistry, 140*(1), 13–21. https://doi.org/10.1093/jb/mvj128.

Virchow, R. (1858). *Die cellular pathologie.* Berlin: August Hirschwald.

Wong, C. W., Lee, A., Shientag, L., Yu, J., Dong, Y., Kao, G., et al. (2001). Apoptosis: An early event in metastatic inefficiency. *Cancer Research, 61*(1), 333–338.

Yoshida, B. A., Sokoloff, M. M., Welch, D. R., & Rinker-Schaeffer, C. W. (2000). Metastasis-suppressor genes: A review and perspective on an emerging field. *Journal of the National Cancer Institute, 92*(21), 1717–1730. https://doi.org/10.1093/jnci/92.21.1717.

Zhang, S., Zhang, N., & Yi, S. (2016). The relationship of lymphatic vessel density, lymphovascular invasion, and lymph node metastasis in breast cancer: A systematic review and meta-analysis. *Oncotarget, 8*, 2863–2873.

CHAPTER 4

Epithelial mesenchymal transition

4.1 Epithelial–mesenchymal transition and mesenchymal to epithelial transition

Epithelial tissues can acquire mesenchymal features during development, tissue repair, wound healing, and cancer invasion (Friedl & Gilmour, 2009). Epithelial–mesenchymal transitions (EMTs) (Fig. 4.1) are classified into three types (Kalluri & Weinberg, 2009): type 1, which occurs during embryonic development that contributes to the formation of the body plan, histogenesis, and organogenesis; type 2, which is associated with adult tissue repair that is involved in cutaneous wound healing, during which keratinocytes migrate across the wound bed to restore the epidermal barrier; moreover, EMT is involved in the development of scarring and fibrosis; type 3, which is involved in cancer progression. The first developmental EMT occurs at gastrulation (Haegel et al., 1995), and a central component of the neural crest cells (Note 1) migration is a programmed EMT (Strobl-Mazzulla & Bronner, 2012; Thiery, Acloque, Huang, & Nieto, 2009).

Approximately 90 per cent of cancers exhibit some degree of EMT during their progression, and epithelial tumors are the result of an EMT process (Thiery et al., 2009). After activation of EMT, tumor cells lose their epithelial features, including cell adhesion and polarity, reorganize their cytoskeleton, and acquire a mesenchymal morphology and the ability to migrate (Thiery et al., 2009). Transforming growth factor beta (TGF-β) is the best known inducer of EMT and acts through Smads to induce EMT-related transcription factors (Xu, Lamouille, & Derynck, 2009). Extracellular matrix (ECM) composition has profound effect on the regulation of EMT via TGF-β availability. αvβ6 integrin engages fibronectin and activates latent TGF-β to induce EMT (Kim et al., 2006), while αvβ6-mediated TGF-β activation can be blocked by inhibitors of Rho

Tumor Microenvironment Regulation of Tumor Expansion
ISBN 978-0-12-822803-6
https://doi.org/10.1016/B978-0-12-822803-6.00002-8

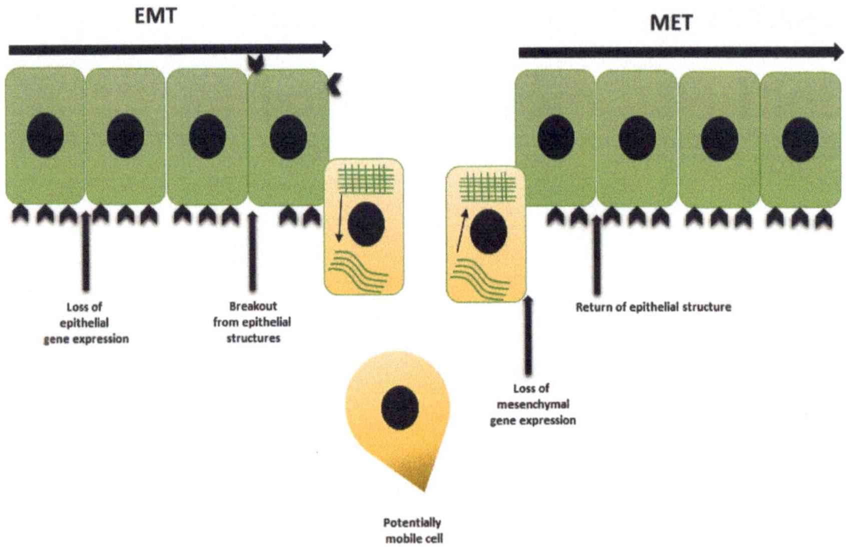

Figure 4.1 Epithelial to mesenchymal transition (EMT) and mesenchymal to epithelial transition (MET).

kinase (ROCK) (Giacomini, Travis, Kudo, & Sheppard, 2012). Moreover, EMT may be induced by hepatocyte growth factor (HGF), epidermal growth factor (EGF), and fibroblast growth factor (FGF) overexpression (Grünert, Jechlinger, & Beug, 2003; Hay, 1995; Huber, Kraut, & Beug, 2005). EGF receptor (EGFR) can be involved in EMT via upregulation of TWIST gene expression (Lo et al., 2007). In the developing lung, FGF acts as a chemoattractant for epithelial cells which express FGF receptors and elongate toward spatial sources of FGF (Warburton et al., 2000). EMT is coordinated by a group of transcription factors, including Snai1/Snail, Snai2/Slug, Twist, and ZEB1, and is characterized by increased expression of the mesenchymal markers vimentin and N-cadherin and downregulation of the E-cadherin gene, an epithelial marker and potent suppressor of tumor cell invasion and metastasis (Lin et al., 2010; Nakamura & Tokura, 2011). Transcriptional repression of E-cadherin by Snail is closely correlated with EMT and the loss of E-cadherin expression is a hallmark of EMT (Batlle et al., 2000; Iwatsuki et al., 2010), and Snail-mutant mice do not survive beyond gastrulation because E-cadherin is unexpressed and cells cannot undergo EMT (Carver, Jiang, Lan, Oram, & Gridley, 2001). Snail is sufficient to induce EMT in tissue culture, and transfection of Snail into epithelial cell lines results in their mesenchymalization associated with a downregulation of E-cadherin expression (Batlle et al., 2000). Notch is involved in the EMT associated with embryonic heart development (Garside, Chang, Karsan, & Hoodless, 2013). EMT is characterized by the breakdown of adherens junctions and loss of epithelial markers, including cytokeratins and E-cadherin (Note 2), and by the overexpression of mesenchymal markers, including fibronectin, N-cadherin, and vimentin, as well as the acquisition of an invasive fibroblastoid phenotype (Garside et al., 2013; Micalizzi, Farabaugh, & Ford, 2010). During the EMT of neural crest formation, a switch from E-cadherin to N-cadherin promotes adherens junction disassembly (Scarpa et al., 2015), and adherens junctions destabilization allows for the release of associated proteins, including β-catenin, which can upregulate EMT genes (Samuel et al., 2011).

The EMT program promotes several important key changes during the initial step of metastasis formation. EMT is responsible for disruption of cell-to-cell contacts, allows cancer cells to become more motile, and enhances the degradation of the ECM and basement membrane, processes essential for local invasion to occur. One of the initial steps of primary tumor invasion is the EMT, during which tumor cells lose epithelial markers and gain mesenchymal traits that confer stemlike properties and a migratory phenotype (Mani et al., 2008). During EMT, epithelial cells exhibit morphological changes: tight junctions, adherens junctions, desmosomes, and gap junctions are lost and the cells undergo actin cytoskeleton reorganization (Fig. 4.2). Decreased E-cadherin function is a component of EMT, invasive tumor growth, and metastasis. Perl, Wilgenbus, Dahl, Semb, and Christofori (1998), using E-cadherin knockout mice and dominant-negative forms of this protein, showed that loss of E-cadherin was associated with pancreatic cell carcinogenesis. The EMT promotes the generation of cancer associated fibroblasts (CAFs) whereby tumor cells of epithelial origin dedifferentiate to generate a mesenchymal-like population that express CAF markers (Petersen et al., 2003). At later stages of metastasis, secondary lesions display an epithelial-like phenotype, suggesting that mesenchymal—epithelial transition (MET) is important for metastatic outgrowth (Note 3) (Fig. 4.1) (Chaffer, Thompson, & Williams, 2007). TGF-β promotes EMT and single cell motility, which enables invasion into blood vessels, while in the absence of TGF-β cells are restricted to collective movement and lymphatic spread (Giampieri, Pinner, & Sahai, 2010). TGF-β is an early tumor suppressor that can subsequently promote

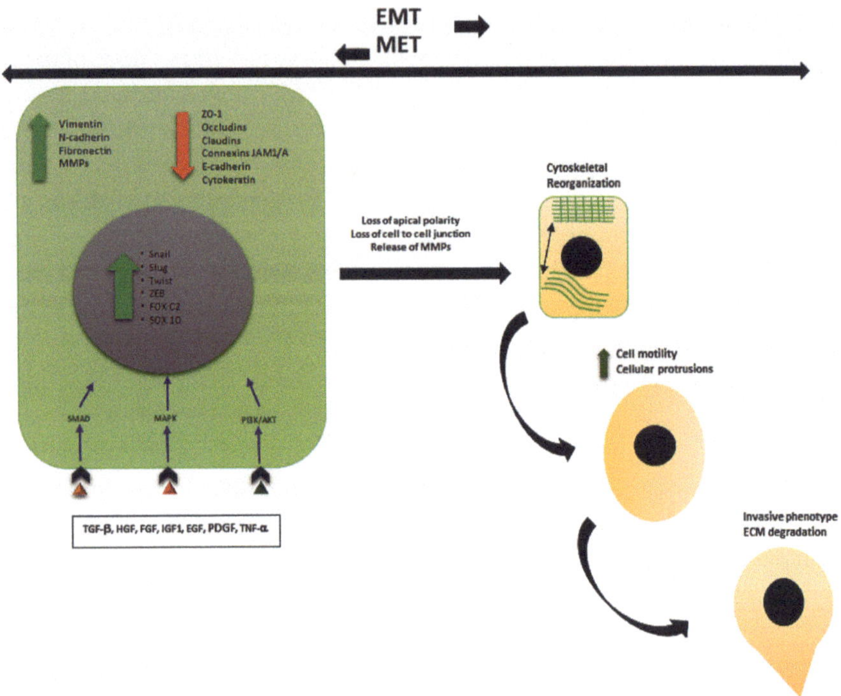

Figure 4.2 Main events in epithelial to mesenchymal transition (EMT).

tumor progression through tumor–stroma interactions resulting in metastasis, immune evasion, and stimulation of angiogenesis (Ikushima & Miyazono, 2010). Macrophages, platelets, and mesenchymal stem cells (MSCs) contribute to EMT at primary sites, allowing tumor cells to separate from neighboring epithelial cell–cell contacts and acquire a mobile and invasive phenotype. During intravasation of tumor cells into the circulation, intravital imaging studies have shown that macrophages are localized to perivascular areas within tumors, where they help tumor cells traverse vessel barriers. Platelets escort tumor cells in the circulation to the site of extravasation, where they bind to areas of vascular retraction and help tumor cells exit the circulation into secondary organs. Myeloid derived suppressor cells (MDSCs) and natural killer (NK) cells populate premetastatic niches, where they help metastatic dissemination by creating a niche permissive to tumor colonization. Fibroblasts upregulate fibronectin, which serves as docking for hematopoietic stem cells (HSCs) and the arrival of tumor cells.

4.2 Notes

1. The neural crest is a stem cell population that gives rise to a vast array of derivatives, ranging from peripheral ganglia to the facial skeleton. This population is induced in the early embryo at the border of the neural plate, which will form the central nervous system (CNS). After neural tube closure, neural crest cells depart from the dorsal CNS via an EMT, forming a migratory mesenchymal cell type that migrates extensively to diverse locations in the embryo.
2. Cell adhesion molecules (CAMs) regulate cell—cell and cell—matrix adhesion and are implicated in almost all stages of metastasis, therefore alterations in normal levels of CAMs, such as E-cadherin, will be significant in tumor progression. E-cadherin is a member of a family of Ca+ dependent CAMs made up of intracellular, extracellular, and transmembrane domains. E-cadherin is located on the lateral basement membrane at the adherent junctions defining apicobasal polarity, connecting neighboring epithelial cells by homotropic calcium-dependent interactions at their extracellular ends, whereas the intracytoplasmic region interacts with other components, mainly β-catenin, thus mediating the mechanical stability of the joints. The alteration of the components of the adhesion bonds, mainly E-cadherin and β-catenin, plays an important role in the induction of the EMT.
3. MET involves morphological changes such as cell polarization and establishment of cell—cell contacts, which are accompanied by changes in the activities of several transcriptional factors that regulate the expression levels of epithelial versus mesenchymal cell markers and signaling molecules. In addition to its occurrence during embryonic developmental programs, MET has been observed during reprogramming of somatic cells into induced pluripotent stem cells and in disease states such as fibrosis, inflammation, and cancers.

References

Batlle, E., Sancho, E., Francí, C., Domínguez, D., Monfar, M., Baulida, J., et al. (2000). The transcription factor Snail is a repressor of E-cadherin gene expression in epithelial tumour cells. *Nature Cell Biology, 2*(2), 84—89. https://doi.org/10.1038/35000034.

Carver, E. A., Jiang, R., Lan, Y., Oram, K. F., & Gridley, T. (2001). The mouse Snail gene encodes a key regulator of the epithelial-mesenchymal transition. *Molecular and Cellular Biology, 21*(23), 8184—8188. https://doi.org/10.1128/MCB.21.23.8184-8188.2001.

Chaffer, C. L., Thompson, E. W., & Williams, E. D. (2007). Mesenchymal to epithelial transition in development and disease. *Cells Tissues Organs, 185*(1—3), 7—19. https://doi.org/10.1159/000101298.

Friedl, P., & Gilmour, D. (2009). Collective cell migration in morphogenesis, regeneration and cancer. *Nature Reviews Molecular Cell Biology, 10*(7), 445–457. https://doi.org/10.1038/nrm2720.

Garside, V. C., Chang, A. C., Karsan, A., & Hoodless, P. A. (2013). Co-ordinating Notch, BMP, and TGF-β signaling during heart valve development. *Cellular and Molecular Life Sciences, 70*(16), 2899–2917. https://doi.org/10.1007/s00018-012-1197-9.

Giacomini, M. M., Travis, M. A., Kudo, M., & Sheppard, D. (2012). Epithelial cells utilize cortical actin/myosin to activate latent TGF-β through integrin αvβ6-dependent physical force. *Experimental Cell Research, 318*(6), 716–722. https://doi.org/10.1016/j.yexcr.2012.01.020.

Giampieri, S., Pinner, S., & Sahai, E. (2010). Intravital imaging illuminates transforming growth factor β signaling switches during metastasis. *Cancer Research, 70*(9), 3435–3439. https://doi.org/10.1158/0008-5472.CAN-10-0466.

Grünert, S., Jechlinger, M., & Beug, H. (2003). Diverse cellular and molecular mechanisms contribute to epithelial plasticity and metastasis. *Nature Reviews Molecular Cell Biology, 4*(8), 657–665. https://doi.org/10.1038/nrm1175.

Haegel, H., Larue, L., Ohsugi, M., Fedorov, L., Herrenknecht, K., & Kemler, R. (1995). Lack of β-catenin affects mouse development at gastrulation. *Development, 121*(11), 3529–3537.

Hay, E. D. (1995). An overview of epithelio-mesenchymal transformation. *Cells Tissues Organs, 154*(1), 8–20. https://doi.org/10.1159/000147748.

Huber, M. A., Kraut, N., & Beug, H. (2005). Molecular requirements for epithelial-mesenchymal transition during tumor progression. *Current Opinion in Cell Biology, 17*(5), 548–558. https://doi.org/10.1016/j.ceb.2005.08.001.

Ikushima, H., & Miyazono, K. (2010). TGFβ signalling: A complex web in cancer progression. *Nature Reviews Cancer, 10*(6), 415–424. https://doi.org/10.1038/nrc2853.

Iwatsuki, M., Mimori, K., Yokobori, T., Ishi, H., Beppu, T., Nakamori, S., et al. (2010). Epithelial-mesenchymal transition in cancer development and its clinical significance. *Cancer Science, 101*(2), 293–299. https://doi.org/10.1111/j.1349-7006.2009.01419.x.

Kalluri, R., & Weinberg, R. A. (2009). The basics of epithelial-mesenchymal transition. *Journal of Clinical Investigation, 119*(6), 1420–1428. https://doi.org/10.1172/JCI39104.

Kim, K. K., Kugler, M. C., Wolters, P. J., Robillard, L., Galvez, M. G., Brumwell, A. M., et al. (2006). Alveolar epithelial cell mesenchymal transition develops in vivo during pulmonary fibrosis and is regulated by the extracellular matrix. *Proceedings of the National Academy of Sciences of the United States of America, 103*(35), 13180–13185. https://doi.org/10.1073/pnas.0605669103.

Lin, K., Baritaki, S., Militello, L., Malaponte, G., Bevelacqua, Y., & Bonavida, B. (2010). The role of B-RAF mutations in melanoma and the induction of EMT via dysregulation of the NF-κB/snail/RKIP/PTEN circuit. *Genes and Cancer, 1*(5), 409–420. https://doi.org/10.1177/1947601910373795.

Lo, H. W., Hsu, S. C., Xia, W., Cao, X., Shih, J. Y., Wei, Y., et al. (2007). Epidermal growth factor receptor cooperates with signal transducer and activator of transcription 3 to induce epithelial-mesenchymal transition in cancer cells via up-regulation of TWIST gene expression. *Cancer Research, 67*(19), 9066–9076. https://doi.org/10.1158/0008-5472.CAN-07-0575.

Mani, S. A., Guo, W., Liao, M.-J., Eaton, E. N., Ayyanan, A., Zhou, A. Y., et al. (2008). The epithelial-mesenchymal transition generates cells with properties of stem cells. *Cell, 133*(4), 704–715. https://doi.org/10.1016/j.cell.2008.03.027.

Micalizzi, D. S., Farabaugh, S. M., & Ford, H. L. (2010). Epithelial-mesenchymal transition in cancer: Parallels between normal development and tumor progression. *Journal of Mammary Gland Biology and Neoplasia, 15*(2), 117–134. https://doi.org/10.1007/s10911-010-9178-9.

Nakamura, M., & Tokura, Y. (2011). Epithelial-mesenchymal transition in the skin. *Journal of Dermatological Science, 61*(1), 7–13. https://doi.org/10.1016/j.jdermsci.2010.11.015

Perl, A.-K., Wilgenbus, P., Dahl, U., Semb, H., & Christofori, G. (1998). A causal role for E-cadherin in the transition from adenoma to carcinoma. *Nature, 392*(6672), 190–193. https://doi.org/10.1038/32433.

Petersen, O. W., Nielsen, H. L., Gudjonsson, T., Villadsen, R., Rank, F., Niebuhr, E., et al. (2003). Epithelial to mesenchymal transition in human breast cancer can provide a nonmalignant stroma. *American Journal of Pathology, 162*(2), 391–402. https://doi.org/10.1016/S0002-9440(10)63834-5.

Samuel, M. S., Lopez, J. I., McGhee, E. J., Croft, D. R., Strachan, D., Timpson, P., et al. (2011). Actomyosin-mediated cellular tension drives increased tissue stiffness and β-catenin activation to induce epidermal hyperplasia and tumor growth. *Cancer Cell, 19*(6), 776–791. https://doi.org/10.1016/j.ccr.2011.05.008.

Scarpa, E., Szabó, A., Bibonne, A., Theveneau, E., Parsons, M., & Mayor, R. (2015). Cadherin switch during EMT in neural crest cells leads to contact inhibition of locomotion via repolarization of forces. *Developmental Cell, 34*(4), 421–434. https://doi.org/10.1016/j.devcel.2015.06.012.

Strobl-Mazzulla, P. H., & Bronner, M. E. (2012). Epithelial to mesenchymal transition: New and old insights from the classical neural crest model. *Seminars in Cancer Biology, 22*(5–6), 411–416. https://doi.org/10.1016/j.semcancer.2012.04.008.

Thiery, J. P., Acloque, H., Huang, R. Y. J., & Nieto, M. A. (2009). Epithelial-mesenchymal transitions in development and disease. *Cell, 139*(5), 871–890. https://doi.org/10.1016/j.cell.2009.11.007.

Warburton, D., Schwarz, M., Tefft, D., Flores-Delgado, G., Anderson, K. D., & Cardoso, W. V. (2000). The molecular basis of lung morphogenesis. *Mechanisms of Development, 92*(1), 55–81. https://doi.org/10.1016/S0925-4773(99)00325-1.

Xu, J., Lamouille, S., & Derynck, R. (2009). TGF-B-induced epithelial to mesenchymal transition. *Cell Research, 19*(2), 156–172. https://doi.org/10.1038/cr.2009.5.

CHAPTER 5
The extracellular matrix

5.1 The components of the extracellular matrix

The extracellular matrix (ECM) is the noncellular component present within all tissues and organs, consisting of a variety of structural and signaling molecules secreted from differentiated mesenchymal cells including chondrocytes and fibroblasts and with biochemical, biomechanical, and structural properties critical for the development of organs. The ECM provides mechanical adhesive support for the cellular constituents, directs their morphological organization, and influences physiological functions, by binding growth factors and interacting with cell-surface receptors. The ECM modulates the growth of tissues to form complex structures that are required for these organs to function and provides structural organization not only through its action as a physical barrier to growing cells but also by activating intracellular signaling. The ECM does this through growth factor distribution modulation, physical anisotropies, and anchorage properties. Moreover, ECM actively undergoes dynamic remodeling during all stages of cancer progression in a complex interplay between cancer cells, resident cells, and noncellular components.

The ECM is composed of a biopolymer fiber network of proteins and proteoglycans that differ in composition and structure between different locations (Fig. 5.1) (Note 1). The ECM forms a mechanically stable support for epithelial cells, allows for the diffusion of oxygen and nutrients between the microvasculature and adjacent tissues, and enables the trafficking of freely moving cells. Through adhesive interactions with integrins expressed on the endothelial cell surface, the ECM orchestrates complex signaling cascades within the cells and affects many fundamental aspects of their biology, including proliferation, migration, cytoskeletal organization, cell shape, survival, and ultimately blood vessel stabilization. The three-dimensional and computational in vitro (Edgar, Sibole, Underwood, Guilkey, & Weiss, 2013; Fang et al., 2013) approach clearly demonstrates that, besides through its remodeling, ECM controls and regulates

Tumor Microenvironment Regulation of Tumor Expansion
ISBN 978-0-12-822803-6
https://doi.org/10.1016/B978-0-12-822803-6.00006-5

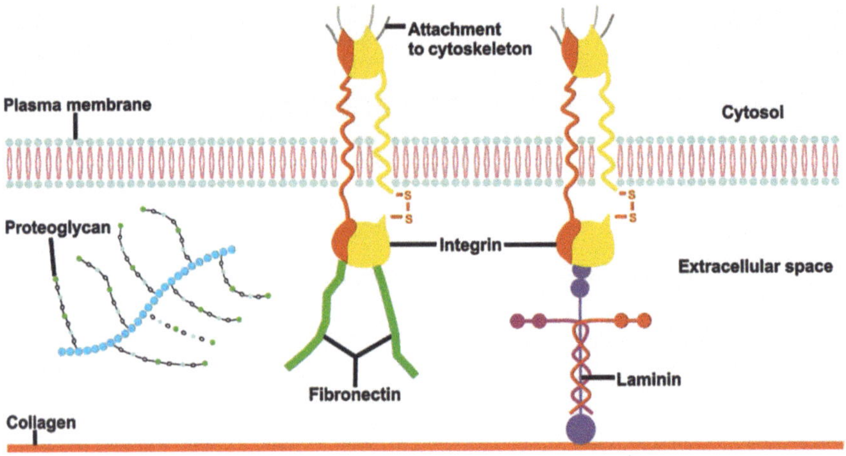

Figure 5.1 ***The general structure of the extracellular matrix (ECM).*** The ECM mainly contained collagen fibers. There are also some glycoproteins acting as adhesion molecules, including fibronectin and laminin. The intracellular part of fibronectin and laminin is highly associated with the cytoskeleton, and thus may promote the anchorage of the cell. Various proteoglycans are also present in the ECM.

physiological and pathological angiogenesis at several levels in several ways. Matrix molecules or fragments show pro- and anti-angiogenic activity are critical in the onset of angiogenesis and angiogenic cytokines directly bind matrix. Fibronectin is a widely distributed glycoprotein and is a component of plasma and of cell surface and ECM in a soluble dimeric and multimeric form, and is localized in ECM underlying endothelial cells. The arginine-glycine—aspartic acid (RGD) motif was the first sequence of fibronectin found to possess cell-adhesive properties (Pierschbacher & Ruoslahti, 1984). The binding of RGD sequence to the integrins α5β1 and αvβ3 in endothelial cells initiates the polymerization of fibronectin locally synthesized (Sottile & Hocking, 2002), which, in turn, regulates cell growth.

Maquart, Siméon, Pasco, and Monboisse (1999) introduced the term "matrikines" to designate peptides produced from degradation of ECM components, as result of enzymatic activity of proteinases produced by connective tissue cells. The same term has been then adopted to describe peptides derived by partial proteolysis of ECM surrounding microvessels. Endothelial cells produce proteinases, whose activation is responsible for the degradation and liberation of matrix fragments which, in turn, can regulate cellular activity by binding specific cell receptors and by activating intracellular signaling pathways. The term "matricryptic sites" has been coined to describe the biologically active sites that are not exposed in the mature, secreted form of ECM molecules but which become exposed after structural or conformational alterations (Davis, Bayless, Davis, & Meininger, 2000). The ECM provides critical and biomechanical cues that direct cell growth, survival, migration, and differentiation, and modulate vascular development and immune function. The ECM is a highly dynamic structure, undergoing continuous remodeling, which consists in the deposition, degradation, and modification of its components. An abnormal ECM dynamic leads to pathological processes including tissue fibrosis and cancer. Vlodavsky, Lui, and Gospodarowicz (1980) demonstrated that ECM determined the morphological appearance, growth pattern, and migratory activity of human tumor cells. Laminin was identified as a major tumor promoter with several active domains, enhancing or inhibiting tumor growth and metastasis (Yamamura, Kibbey, & Kleinman, 1993). ECM remodeling, including deposition, reorganization, matrix metalloproteinases (MMPs) activity, contributes to tumor progression and plays an integral role in the regulation of tumor cell function and treatment responsiveness (Malik, Lelkes, & Cukierman, 2015).

5.2 Proteolytic enzymes

Two classes of proteolytic enzymes appear to be mainly involved in the matrix degradation: plasminogen activator (PA)/plasmin system and matrix metalloproteinases (MMPs) (Note 2). The PA/plasmin system is an enzymatic cascade involved in the control of fibrin degradation, matrix turnover, and cell invasion. The physiological activators urokinase-type PA or tissue-type PA mediate the conversion of the inactive plasma zymogen, plasminogen, to the serine protease plasmin. The latter belongs to the large serine proteinase family and can act directly or indirectly by cleaving numerous ECM proteins, including fibronectin, laminin, thrombospondin, and von Willebrand factor, by activating MMPs (Davis et al., 2000) or by liberating growth factors and cytokines sequestered within the ECM (Saksela, Moscatelli, Sommer, & Rifkin, 1988).

Some endogenous proteins or fragments of proteins formed in the body act as physiological inhibitors of angiogenesis (Ribatti, 2009). Many of Matrikines and Matricryptic sites, resulted from PA/plasmin system and MMPs activity, are important physiologic angiogenesis inhibitors. Among these, endostatin has been extensively studied. It is a proteolytic fragment of the C-terminal noncollagenous domain of collagen XVIII, isolated from the conditioned media of a nonmetastatic murine hemangioendothelioma cell line (O'Reilly et al., 1997) Its role as a local inhibitor of angiogenesis has been demonstrated in vitro and in vivo (O'Reilly et al., 1997) (Note 3). Tumstatin, canstatin, and arresten are derived from degradation of $\alpha3$, $\alpha2$, and $\alpha1$ chain, respectively, of type IV collagen (Boosani, Nalabothula, Sheibani, & Sudhakar, 2010; Hamano et al., 2003; Kamphaus et al., 2000). Tetrastatin, pentastatin, and hexastatin are derived from degradation of type IV collagen, from $\alpha4$, $\alpha5$, and $\alpha6$ chain, respectively (Karagiannis & Popel, 2007; Mundel, Yliniemi, & Maeshima, 1982; Petitclerc et al., 2000). Endorepellin derives from perlecan (Mongiat, Sweeney, San Antonio, Fu, & Iozzo, 2003) and affects angiogenesis by interacting with the vascular endothelial growth factor receptor (VEGFR) through its laminin G-like, leading to the downstream of VEGF signaling (Goyal et al., 2012). Recombinant tumstatin is antiangiogenic in several in vitro and in vivo experimental models (Maeshima et al., 2002), and arresten inhibits growth of primary tumors and metastases in mouse xenograft models (Sudhakar et al., 2005). VEGF bounds the cell surface or ECM and various MMPs and PA (Keyt et al., 1996; Lee, Jilan, Nikolova, Carpizo, & Luisa Iruela-Arispe, 2005) generate diffusible, nonheparin-binding fragments.

5.3 Angiogenesis stimulators and inhibitors in the extracellular matrix

Alternative mRNA splicing of the VEGF-A transcript gives rise to three major isoforms with different degrees of affinity for the ECM (Tischer et al., 1991), including 120, 164, and 188 residues in rodents (or 121, 165, and 189 in humans, respectively) due to the presence of heparin binding domains that interact with ECM proteoglycans, so that the matrix affinity is very low in the shortest isoform and increases with molecular size (Ferrara, 2010). VEGF 164/165 binds the coreceptor neuropilin-1 (NRP-1), enhancing activation of VEGFR-2 and endothelial cell proliferation and migration, whereas VEGF 120/121 does not (Lampropoulou & Ruhrberg, 2014). A distal splice site in the last exon of VEGF genes gives rise to a set of antiangiogenic variants (Harper & Bates, 2008). As a consequence of differential matrix binding, VEGF 120/121 is highly diffusible in tissues, whereas VEGF 188/189 remains localized at the site of secretion and VEGF 164/165 generates intermediate gradients of concentration around the producing cells. Two classes of VEGF binding sites have been identified on fibronectin: one constitutively available and the other whose availability is modulated by the conformational state of fibronectin, which, in turn, depends on heparin interaction (Mitsi, Hong, Costello, & Nugent, 2006; Mitsi, Forsten-Williams, Gopalakrishnan, & Nugent, 2008).

Fibroblast growth factor-2 (FGF-2) binds with high affinity to heparan sulfate proteoglycans located on the surface of most cells and within the ECM (Ishai-Michaeli et al., 1992). Heparan sulfate proteoglycans, particularly perlecan, modulate the binding of FGF-2 to its specific tyrosine kinase receptors (Aviezer et al., 1994) and protect the growth factor from proteolytic degradation by extracellular proteinases (Saksela & Rifkin, 1990). The activity of transforming growth factor beta-1 (TGF-β1) is strictly dependent on composition and organization of the ECM: TGF-β1 treatment elicits the formation of calcium and magnesium dependent tubelike structures, mimicking angiogenesis, in three-dimensional cultures, whereas it induces endothelial cells apoptosis in two-dimensional cultures (Merwin, Anderson, Kocher, Van Itallie, & Madri, 1990). Angiopoietin-1 (Ang-1) is secreted and incorporated into and sequestered by the ECM. Unlike Ang-1, its antagonist, Ang-2, is not incorporated into ECM and inhibits blood vessel formation (Maisonpierre et al., 1997).

5.4 Desmoplasia and cancer

Desmoplasia, the formation of a dense ECM, is associated with several forms of cancer. It is characterized by extensive proliferation of myofibroblasts expressing α smooth actin in the ECM (Liu et al., 2012) and by an increase in fibrillary collagen, fibronectin, proteoglycans, and tenascin (Kalluri & Zeisberg, 2006). The tumor-associated ECM is characterized by altered posttranslational modifications, such as cross-linking, leading to the generation of a dense fibrotic protumoringenic microenvironment (Cox & Erler, 2014). Desmoplasia prevents tumor growth (Dvorak, Dvorak, Manseau, Pyne, & Dlckersln, 1981) or is associated with more aggressive and invasive cancer (Kalluri & Zeisberg, 2006) (Note 4). A collagen-rich ECM promotes macrophage proliferation and activation (Wesley, Meng, Godin, & Galis, 1998) and favors a protumorigenic M2 polarization phenotype, whereas a fibronectin-rich ECM promotes the M1 or antitumorigenic potential of macrophages (Aviezer et al., 1994; Perri et al., 1982; Stahl et al., 2013). A correlation between fibrotic stroma and cancer is well established in the liver (Maher & Bissell, 1993) and increased stromal density correlates with a higher likelihood of developing breast cancer

(Wolfe, 1976). Gene-expressing profiling has shown that genes encoding for ECM components are dysregulated during tumor progression (Eckhardt et al., 2005).

5.5 Notes

1. Proteoglycans are characterized as proteins that have glycosaminoglycans (GAGs) covalently bonded to them. These GAGs are long chains of negatively charged disaccharide repeats that can either be heparin sulfate, chondroitin/dermatan sulfate, hyaluronan, or keratin sulfate. Due to the negative charge of these GAGs, proteoglycans are able to sequester water and cations, which gives them their space-filling and lubrication functions. Proteoglycans vary in form and structure that confer them different functions in the ECM. They are integral in the maintenance of a healthy ECM without which would lead to a nonfunctional ECM and a collapse of its structure.
2. The plasminogen activator/plasmin system is an enzymatic cascade involved in the control of fibrin degradation, matrix turnover, and cell invasion. Plasmin is, in turn, able to degrade fibronectin, laminin, vitronectin, proteoglycans, as well as fibrin and activate latent collagenases. Currently there are 28 known MMPs identified in vertebrates, including 23 in humans. Common classification distinguishes a number of metalloproteinases, starting with MMP-1 and ending with MMP-28, but it does not include MMP-4, MMP-5, MMP-6, MMP-22, since these enzymes were discovered simultaneously by different research teams. MMPs are involved in trophoblast implantation, embryogenesis, bone growth, angiogenesis, wound healing, and tissue regeneration. The expression of MMPs' genes is observed in the connective tissue cells, primarily in fibroblasts, but also in neutrophils, monocytes, macrophages, and endothelial cells.
3. Endostatin was first isolated in Folkman's lab in 1997, and was reported to dramatically shrink tumor blood formation. Its insoluble and unstable nature coupled with the high cost of synthesizing the endostatin protein doomed it for clinical cancer treatment. Chinese scientists found a way to refold the protein, making it cost-effective to manufacture a recombinant human endostatin, a soluble and stable form of endostatin. A number of clinical studies have demonstrated the significant survival benefit of rh-endostatin in treating late-stage non small cell lung cancer (NSCLC) and, as a result, rh-endostatin (Endostar) was approved by the

State Food and Drug Administration of China (CFDA) in September 2005 as a treatment option for NSCLC. Since then, increasing bodies of clinical data and experience have been obtained from a variety of other different cancers, such as malignant serous effusion, melanoma, colon cancer, gastric cancer, breast cancer, nasopharyngeal cancers.

4. The term desmoplasia (from the Greek word desmos, *to fetter or restrain*; and plasis, *formation*), has been used by pathologists for more than a century in reference to the formation of excessive connective tissue around invasive carcinoma. Desmoplastic tissues are characterized by alterations of the tumor stroma that can range from an abundance of cellular elements such as fibroblasts, vascular cells, and immune cells with little ECM to the presence of an abundant collagen-rich ECM with few fibroblasts and myofibroblasts. Desmoplasia is the result of an increased synthesis of ECM proteins and collagen by stromal cells such as myofibroblasts. A dense ECM represents a barrier against cancer invasion and metastasis. It was therefore considered to be a reaction and response of the host tissue against invasive cancer cells.

References

Aviezer, D., Levy, E., Safran, M., Svahn, C., Buddecke, E., Schmidt, A., et al. (1994). Differential structural requirements of heparin and heparan sulfate proteoglycans that promote binding of basic fibroblast growth factor to its receptor. *Journal of Biological Chemistry, 269*(1), 114–121.

Boosani, C. S., Nalabothula, N., Sheibani, N., & Sudhakar, A. (2010). Inhibitory effects of arresten on bFGF-induced proliferation, migration, and matrix metalloproteinase-2 activation in mouse retinal endothelial cells impact of arresten on mouse retinal Endothelial Cells. *Current Eye Research, 35*(1), 45–55. https://doi.org/10.3109/02713680903374208.

Cox, T. R., & Erler, J. T. (2014). Molecular pathways: Connecting fibrosis and solid tumor metastasis. *Clinical Cancer Research, 20*(14), 3637–3643. https://doi.org/10.1158/1078-0432.CCR-13-1059.

Davis, G. E., Bayless, K. J., Davis, M. J., & Meininger, G. A. (2000). Regulation of tissue injury responses by the exposure of matricryptic sites within extracellular matrix molecules. *American Journal of Pathology, 156*(5), 1489–1498. https://doi.org/10.1016/S0002-9440(10)65020-1.

Dvorak, H. F., Dvorak, A. M., Manseau, E. J., Pyne, K., & Dlckersln, G. R. (1981). Human breast carcinoma: Fibrin deposits ans desmoplasia. Inflammatory cell type and distribution. Microvasculature and infarction. *Journal of the National Cancer Institute, 67*(2), 335–345. https://doi.org/10.1093/jnci/67.2.335.

Eckhardt, B. L., Parker, B. S., Van Laar, R. K., Restall, C. M., Natoli, A. L., Tavaria, M. D., et al. (2005). Genomic analysis of a spontaneous model of breast cancer metastasis to bone reveals a role for the extracellular matrix. *Molecular Cancer Research, 3*(1), 1–13.

Edgar, L. T., Sibole, S. C., Underwood, C. J., Guilkey, J. E., & Weiss, J. A. (2013). A computational model of in vitro angiogenesis based on extracellular matrix fibre orientation. *Computer Methods in Biomechanics and Biomedical Engineering, 16*(7), 790–801. https://doi.org/10.1080/10255842.2012.662678.

Fang, C., Man, Y.-G., Cuttitta, F., Stetler-Stevenson, W., Salomon, D., Mazar, A., et al. (2013). Novel phenotypic fluorescent three-dimensional co-culture platforms for recapitulating tumor in vivo progression and for personalized therapy. *Journal of Cancer, 4*(9), 755–763. https://doi.org/10.7150/jca.7813.

Ferrara, N. (2010). Binding to the extracellular matrix and proteolytic processing: Two key mechanisms regulating vascular endothelial growth factor action. *Molecular Biology of the Cell, 21*(5), 687–690. https://doi.org/10.1091/mbc.E09-07-0590.

Goyal, A., Poluzzi, C., Willis, C. D., Smythies, J., Shellard, A., Neill, T., et al. (2012). Endorepellin affects angiogenesis by antagonizing diverse vascular endothelial growth factor receptor 2 (VEGFR2)-evoked signaling pathways: Transcriptional repression of hypoxia-inducible factor 1α and VEGFA and concurrent inhibition of nuclear factor of activated T cell 1 (NFAT1) activation. *Journal of Biological Chemistry, 287*(52), 43543–43556. https://doi.org/10.1074/jbc.M112.401786.

Hamano, Y., Zeisberg, M., Sugimoto, H., Lively, J. C., Maeshima, Y., Yang, C., et al. (2003). Physiological levels of tumstatin, a fragment of collagen IV α3 chain, are generated by MMP-9 proteolysis and suppress angiogenesis via αVβ3 integrin. *Cancer Cell, 3*(6), 589–601. https://doi.org/10.1016/S1535-6108(03)00133-8.

Harper, S. J., & Bates, D. O. (2008). VEGF-A splicing: The key to anti-angiogenic therapeutics? *Nature Reviews Cancer, 8*(11), 880–887. https://doi.org/10.1038/nrc2505.

Ishai-Michaeli, R., Korner, G., Vlodavsky, I., Svahn, C. M., Weber, M., Ekre, H. P., et al. (1992). Importance of size and sulfation of heparin in release of basic fibroblast growth factor from the vascular endothelium and extracellular matrix. *Biochemistry, 31*(7), 2080–2088. https://doi.org/10.1021/bi00122a027.

Kalluri, R., & Zeisberg, M. (2006). Fibroblasts in cancer. *Nature Reviews Cancer, 6*(5), 392–401. https://doi.org/10.1038/nrc1877.

Kamphaus, G. D., Colorado, P. C., Panka, D. J., Hopfer, H., Ramchandran, R., Torre, A., et al. (2000). Canstatin, a novel matrix-derived inhibitor of angiogenesis and tumor growth. *Journal of Biological Chemistry, 275*(2), 1209–1215. https://doi.org/10.1074/jbc.275.2.1209.

Karagiannis, E. D., & Popel, A. S. (2007). Identification of novel short peptides derived from the alpha 4, alpha 5, and alpha 6 fibrils of type IV collagen with anti-angiogenic properties. *Biochemical and Biophysical Research Communications, 354*(2), 434–439. https://doi.org/10.1016/j.bbrc.2006.12.231.

Keyt, B. A., Berleau, L. T., Nguyen, H. V., Chen, H., Heinsohn, H., Vandlen, R., et al. (1996). The carboxyl-terminal domain (111-165) of vascular endothelial growth factor is critical for its mitogenic potency. *Journal of Biological Chemistry, 271*(13), 7788–7795. https://doi.org/10.1074/jbc.271.13.7788.

Lampropoulou, A., & Ruhrberg, C. (2014). Neuropilin regulation of angiogenesis. *Biochemical Society Transactions, 42*(6), 1623–1628. https://doi.org/10.1042/bst20140244.

Lee, S., Jilan, S. M., Nikolova, G. V., Carpizo, D., & Luisa Iruela-Arispe, M. (2005). Processing of VEGF-A by matrix metalloproteinases regulates bioavailability and vascular patterning in tumors. *Journal of Cell Biology, 169*(4), 681–691. https://doi.org/10.1083/jcb.200409115.

Liu, H., Ma, Q., Xu, Q., Lei, J., Li, X., Wang, Z., et al. (2012). Therapeutic potential of perineural invasion, hypoxia and desmoplasia in pancreatic cancer. *Current Pharmaceutical Design, 18*(17), 2395–2403. https://doi.org/10.2174/13816128112092395.

Maeshima, Y., Sudhakar, A., Lively, J. C., Ueki, K., Kharbanda, S., Kahn, C. R., et al. (2002). Tumstatin, an endothelial cell-specific inhibitor of protein synthesis. *Science, 295*(5552), 140–143. https://doi.org/10.1126/science.1065298.

Maher, J. J., & Bissell, D. M. (1993). Cell-matrix interactions in liver. *Seminars in Cell Biology, 4*(3), 189–201. https://doi.org/10.1006/scel.1993.1023.

Maisonpierre, P. C., Suri, C., Jones, P. F., Bartunkova, S., Wiegand, S. J., Radziejewski, C., et al. (1997). Angiopoietin-2, a natural antagonist for Tie2 that disrupts in vivo angiogenesis. *Science, 277*(5322), 55–60. https://doi.org/10.1126/science.277.5322.55.

Malik, R., Lelkes, P. I., & Cukierman, E. (2015). Biomechanical and biochemical remodeling of stromal extracellular matrix in cancer. *Trends in Biotechnology, 33*(4), 230–236. https://doi.org/10.1016/j.tibtech.2015.01.004.

Maquart, F. X., Siméon, A., Pasco, S., & Monboisse, J. C. (1999). Régulation de l'activité cellulaire par la matrice extracelulaire: Le concept de matrikines. *Journal de la Société de biologie, 193*(4–5), 423–428.

Merwin, J. R., Anderson, J. M., Kocher, O., Van Itallie, C. M., & Madri, J. A. (1990). Transforming growth factor beta1 modulates extracellular matrix organization and cell-cell junctional complex formation during in vitro angiogenesis. *Journal of Cellular Physiology, 142*(1), 117–128. https://doi.org/10.1002/jcp.1041420115.

Mitsi, M., Forsten-Williams, K., Gopalakrishnan, M., & Nugent, M. A. (2008). A catalytic role of heparin within the extracellular matrix. *Journal of Biological Chemistry, 283*(50), 34796–34807. https://doi.org/10.1074/jbc.M806692200.

Mitsi, M., Hong, Z., Costello, C. E., & Nugent, M. A. (2006). Heparin-mediated conformational changes in fibronectin expose vascular endothelial growth factor binding sites. *Biochemistry, 45*(34), 10319–10328. https://doi.org/10.1021/bi060974p.

Mongiat, M., Sweeney, S. M., San Antonio, J. D., Fu, J., & Iozzo, R. V. (2003). Endorepellin, a novel inhibitor of angiogenesis derived from the C terminus of perlecan. *Journal of Biological Chemistry, 278*(6), 4238–4249. https://doi.org/10.1074/jbc.M210445200.

Mundel, Yliniemi, & Maeshima. (1982). Type IV collagen alpha6 chain-derived non-collagenous domain 1 (alpha6(IV)NC1) inhibits angiogenesis and tumor growth. *International Journal of Cancer, 122*, 430–435.

O'Reilly, M. S., Boehm, T., Shing, Y., Fukai, N., Vasios, G., Lane, W. S., et al. (1997). Endostatin: An endogenous inhibitor of angiogenesis and tumor growth. *Cell, 88*(2), 277–285. https://doi.org/10.1016/s0092-8674(00)81848-6.

Perri, R. T., Kay, N. E., McCarthy, J., Vessella, R. L., Jacob, H. S., & Furcht, L. T. (1982). Fibronectin enhances in vitro monocyte-macrophage-mediated tumoricidal activity. *Blood, 60*(2), 430–435. https://doi.org/10.1182/blood.v60.2.430.430.

Petitclerc, E., Boutaud, A., Prestayko, A., Xu, J., Sado, Y., Ninomiya, Y., et al. (2000). New functions for non-collagenous domains of human collagen type IV. Novel integrin ligands inhibiting angiogenesis and tumor growth in vivo. *Journal of Biological Chemistry, 275*(11), 8051–8061. https://doi.org/10.1074/jbc.275.11.8051.

Pierschbacher, M. D., & Ruoslahti, E. (1984). Cell attachment activity of fibronectin can be duplicated by small synthetic fragments of the molecule. *Nature, 309*(5963), 30–33. https://doi.org/10.1038/309030a0.

Ribatti, D. (2009). Endogenous inhibitors of angiogenesis. A historical review. *Leukemia Research, 33*, 638–644.

Saksela, O., Moscatelli, D., Sommer, A., & Rifkin, D. B. (1988). Endothelial cell-derived heparan sulfate binds basic fibroblast growth factor and protects it from proteolytic degradation. *Journal of Cell Biology, 107*(2), 743–751. https://doi.org/10.1083/jcb.107.2.743.

Saksela, O., & Rifkin, D. B. (1990). Release of basic fibroblast growth factor-heparan sulfate complexes from endothelial cells by plasminogen activator-mediated proteolytic activity. *Journal of Cell Biology, 110*(3), 767–775. https://doi.org/10.1083/jcb.110.3.767.

Sottile, J., & Hocking, D. C. (2002). Fibronectin polymerization regulates the composition and stability of extracellular matrix fibrils and cell-matrix adhesions. *Molecular Biology of the Cell, 13*(10), 3546–3559. https://doi.org/10.1091/mbc.E02-01-0048.

Stahl, M., Schupp, J., Jäger, B., Schmid, M., Zissel, G., Müller-Quernheim, J., et al. (2013). Lung collagens perpetuate pulmonary fibrosis via CD204 and M2 macrophage activation. *PLoS One, 8*(11), e81382. https://doi.org/10.1371/journal.pone.0081382.

Sudhakar, A., Nyberg, P., Keshamouni, V. G., Mannam, A. P., Li, J., Sugimoto, H., et al. (2005). Human α1 type IV collagen NC1 domain exhibits distinct antiangiogenic activity mediated by α1β1 integrin. *Journal of Clinical Investigation, 115*(10), 2801–2810. https://doi.org/10.1172/JCI24813.

Tischer, E., Mitchell, R., Hartman, T., Silva, M., Gospodarowicz, D., Fiddes, J. C., et al. (1991). The human gene for vascular endothelial growth factor. Multiple protein forms are encoded through alternative exon splicing. *Journal of Biological Chemistry, 266*(18), 11947–11954. https://www.ncbi.nlm.nih.gov/pubmed/1711045.

Vlodavsky, I., Lui, G., & Gospodarowicz, D. (1980). Morphological appearance, growth behavior and migratory activity of human tumor cells maintained on extracellular matrix versus plastic. *Cell, 19*(3), 607–616. https://doi.org/10.1016/s0092-8674(80)80037-7.

Wesley, R. B., 2nd, Meng, X., Godin, D., & Galis, Z. S. (1998). Extracellular matrix modulates macrophage functions characteristic to atheroma: collagen type I enhances acquisition of resident macrophage traits by human peripheral blood monocytes in vitro. *Arteriosclerosis, Thrombosis, and Vascular Biology, 18*(3), 432–440. https://doi.org/10.1161/01.atv.18.3.432.

Wolfe, J. N. (1976). Risk for breast cancer development determined by mammographic parenchymal pattern. *Cancer, 37*(5), 2486–2492. https://doi.org/10.1002/1097-0142(197605)37:5<2486::AID-CNCR2820370542>3.0.CO;2-8.

Yamamura, K., Kibbey, M. C., & Kleinman, H. K. (1993). Melanoma cells selected for adhesion to laminin peptides have different malignant properties. *Cancer Research, 53*(2), 423–428.

CHAPTER 6

Tumor blood vessels and tumor endothelial cells

6.1 Tumor blood vessels

In 1908, Goldmann described in his work "The Growth of Malignant Disease in Man and the Lower Animals, with special reference to the Vascular System" tumor blood vessels as follows: "The normal blood vessels of the organs in which the tumor is developing are disturbed by chaotic growth, there is a dilatation and spiraling of the affected vessels, marked capillary budding and new vessel formation, particularly at the advanced border" (Goldmann, 1908). Tumor blood vessels are irregular in size, shape, and branching pattern, lack the normal hierarchy, and do not display the recognizable features of arterioles, capillaries, or venules (Table 6.1). The arrangement of vessels in the center of a tumor is much more chaotic than at its edges. Their haphazard branching patterns and larger, less regular diameters contribute to the nonuniform perfusion of tumor cells.

Tumor vessel leakiness and compression create a vicious cycle responsible for both acute and chronic hypoxia. The functional and structural abnormalities of tumor blood vessels together with the unresponsiveness of the endothelium to inflammatory stimuli caused by proangiogenic factors decrease the recruitment of immune effector cells into the tumor, thus limiting the effectiveness of cancer immunotherapies that contributes to the tumor immune escape and severely impairs conventional cancer therapies (Mai et al., 2013). Abnormalities are present in all components of the vessels wall of tumors. The organ- and tissue-specific vascular architecture are not retained. Even large caliber vessels are mainly composed of just an endothelium and a basement membrane. Vessel density is heterogeneous: the highest values are usually found in the invading tumor edge, where the density is 4 to 10 times greater than inside the tumor (Giatromanolaki et al., 2002).

Tumor Microenvironment Regulation of Tumor Expansion
ISBN 978-0-12-822803-6
https://doi.org/10.1016/B978-0-12-822803-6.00005-3

Table 6.1 Principal differences between tumor and normal blood vessels.

Tumor blood vessels
Excessive branching
Abnormal basement membrane
Loose association between pericytes and endothelial cells
Lack of arteriole-capillary-venule hierarchy
Chaotic flow patterns
Normal blood vessels
Recognizable hierarchy
Pericytes are closely associated with endothelial cells
Pericytes cover much of vessel surface

6.2 Tumor endothelial cells

Endothelial cells of mature, quiescent vessels are characteristically low proliferative and their estimated turnover times are measured in years, whereas those of tumor vessels are markedly dependent on growth factors for survival. Vascular endothelial growth factor (VEGF) has been assigned a central role in the induction of host vessels into a growing tumor. Endothelial cells contribute at least in two ways to tumor progression; first, recruitment of endothelial precursor cells (EPCs) stimulates the sprouting of new tumor blood vessels; second, changes in the structure of the existing vascular endothelium create an escape route for cancer cells to generate distant metastases. When endothelial cells invade a newly formed tumor, they come into contact with tumor cells that produce VEGF, which may be responsible not only for vascular proliferation but also for the altered permeability of the newly formed vessels. Moreover, extracellular vesicles released from tumor cells, cancer associated fibroblasts (CAFs), and tumor associated macrophages (TAMs) stimulate angiogenesis. Tumors can also promote lymphangiogenesis, through the secretion of VEGF-C, VEGF-D, interleukin-1 beta (IL-1β), and fibroblast growth factor-2 (FGF-2), which in turn promotes cancer dissemination (Karaman & Detmar, 2014). Furthermore, it results in extravasation of plasma proteins and even erythrocytes and may facilitate the traffic of tumor cells into the bloodstream and the formation of metastases (Dvorak, Nagy, Dvorak, & Dvorak, 1988).

Leakiness has been attributed to highly active angiogenesis and microvascular remodeling, but its structural basis and mechanism are unclear.

Intercellular gaps, transendothelial holes, vesiculo-vacuolar organelles (VVOs) traverse endothelial cells from lumen to albumen and additionally open to the interendothelial cell clefts, and endothelial fenestrae are present in the endothelium of tumor vessels (Dvorak et al., 1988). Vessel leakiness reduces blood flow by increasing hematocrit and blood viscosity and consequent reduction of blood flow rate (Stylianopoulos, Munn, & Jain, 2018). The phage display library technique has been successfully used to discover tumor cell surface binding peptides. A panel of peptide motifs, including the sequences CDCRGDCFC (termed RGD-4C), NGR, CPRECES, and GSL, have been assembled that target the tumor blood vessels (Kolonin, Pasqualini, & Arap, 2001). The order of blood flow in the normal vessels is from arteries to arterioles, and subsequently to capillaries, postcapillary venules, and lastly veins. Endothelial cells of normal arteries, capillaries, and veins exhibit differential gene expression and molecular identities already prior to the onset of blood flow (Langenkamp & Molema, 2009), indicating the importance of hitherto unidentified microenvironmental factors (Note 1). Tumor endothelial cells proliferate 50–200 times faster than normal endothelial cells (Vermeulen et al., 1995). Their constant proliferation rate reflects the angiogenesis (Table 6.2) that accompanies an increase in tumor volume, whereas in other regions they undergo apoptosis in parallel with tumor necrosis and vessel regression.

The tumor-associated endothelium is structurally defective. Discontinuities or gaps (<2 μm in diameter) that allow hemorrhage and facilitate permeability are common features (Fig. 6.1). Cell contacts are usually

Table 6.2 Angiocrine factors produced by tumor endothelial cells.

Angiopoietin-2 (Ang-2)
Bone morphogenetic protein-2, Bone morphogenetic protein-4 (BMP-2, BMP-4)
C-X-C motif chemokine ligand 12 (CXCL-12)
Fibroblast growth factor-2 (FGF-2)
Granulocyte colony stimulating factor (G-CSF)
Granulocyte macrophage colony stimulating factor (GM-CSF)
Insulin like growth factor-1 (ILGF-1)
Interleukin-6, Interleukin-8, (IL-6, -8)
Pentraxin-3 (PTX-3)
Placental growth factor-1 (PlGF-1)
Platelet derived growth factor (PDGF)
Transforming growth factor beta (TGF-β)
Vascular endothelial growth factor (VEGF)

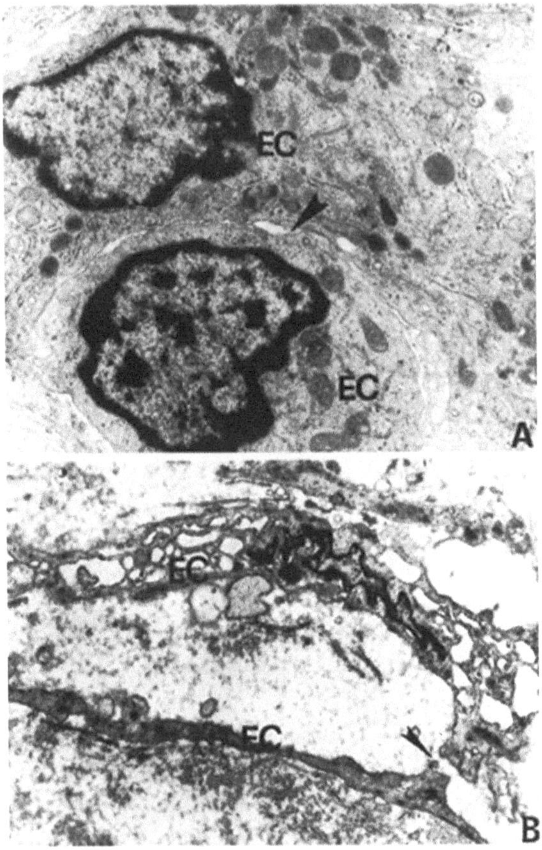

Figure 6.1 Ultrastructural finding in the stroma of high-grade lymphoma. A new capillary with a slit-like lumen (arrowhead), formed by two endothelial cells arranged in parallel. B. Ultrastructural finding in the stroma of a low-grade lymphoma. A fenestrated capillary, showing flat endothelial cells (EC) with a fenestra (arrowhead). Original magnification: (A) ×16.200; (B) ×4800. *(Reproduced from Ribatti et al., 1996.)*

poorly differentiated and no complex contact structures exist. Defects in endothelial cell barrier function, due to abnormal cell–cell junctions and other changes, exaggerate leakiness (Table 6.3). This correlates with histological grade and malignant potential (Daldrup et al., 1998) and can be exploited in locating tumors by imaging contrast media and in the delivery of macromolecular therapeutics (McDonald & Choyke, 2003). Tumor angiogenesis depends on the metabolism of intratumor endothelial cells, which can be more dependent on glycolysis than endothelial cells in healthy organs. Nonetheless, cancer cell hyperglycolysis produces excessive

Table 6.3 Factors implicated in altering vascular permeability.

Vascular endothelial growth factor (VEGF)
A disintegrin and metalloproteinase domain-containing proteins 12 (ADAM 12)
Epiregulin
Cyclooxygenase-2 (COX-2)
Matrix metalloproteinases-2, -3, -9, -10 (MMPs-2, -3,-9, -10)
Angiopoietin-4 (Ang-4)
Angiotensin-2
Stromal derived factor-1 (SDF-1)

lactate, which is consumed by endothelial cells and induces hypoxia inducible factor-1 alpha (HIF-1α) activation, thereby promoting angiogenesis (Cantelmo et al., 2016). Tumor endothelial cells also differ for the profile and level of cell adhesion molecule they express. They preferentially overexpress the cell surface molecules that are absent in quiescent blood vessels, including integrin αvβ3 and αvβ5, the fibronectin extra domain B (EBD), E-selectin, endoglin, endosialin, VEGF receptors (VEGFRs), and prostate specific membrane antigen (PSMA), all of which stimulate endothelium adhesion and migration (Magnussen et al., 2005; Thorpe, 2004). Tumors treated with antagonists of integrin αv β3 display increased apoptosis and vessel regression. One feature of the role of integrin αv β3 is that it mediates endothelial cell adhesion to extracellular matrix (ECM) molecules. Vascular endothelial cadherin (VE-cadherin) is poorly expressed in tumor vessels. It results in their destabilization and may lead to abnormal remodeling. Endothelial cells lining tumor blood vessels express several cell surface markers that are absent in quiescent blood vessels. Ligand-directed vascular targeting can be accomplished by antibodies, specific peptides or growth factors complexed with immunomodulatory, procoagulant or cytotoxic molecules (Thorpe, 2004).

VEGF and its receptors are upregulated in tumor microenvironment, leading to a high concentration of occupied receptors on tumor vascular endothelium. VEGF immediate, short-term (minutes) activity consists in increase of microvascular permeability to plasma and plasma proteins with a potency some 50,000 times that of histamine (Senger, Perruzzi, Feder, & Dvorak, 1986). In the middle to longer terms (days, weeks) VEGF induces endothelial cell proliferation and survival (Ribatti, 2005). VEGFR complexes specifically target on tumor endothelium for antibodies in vivo. Targeting VEGF or its receptors with monoclonal antibodies (such as bevacizumab) or small molecule inhibitors of VEGFR tyrosine kinase has confirmed the anticancer activity of these agents (Ferrara, 2002). Fusion proteins and chemical conjugates of VEGF and diphtheria toxin or gelonin induced tumor regression in mice (Thorpe, 2004). Retroviruses have been engineered so that they can be coated with an antibody (e.g., anti-VEGFR-2) for the selective delivery of genes to tumor endothelium (Masood et al., 2001). Tumor endothelial cells have an abnormal karyotype characterized by various chromosomal abnormalities and aneuploidy. The genetic changes lead to the upregulated expression of proangiogenic genes and angiocrine factors. Furthermore, the upregulation of stemness genes such as stem cell antigen-1 (Sca-1) (Matsuda et al., 2010), Multi-Drug Resistance Gene 1 (MDR1) (Akiyama et al., 2012) and aldehyde dehydrogenase (ALDH) (Hida et al., 2017) leads to the development of drug resistant phenotype in the tumor endothelial cells. Croix (Croix, 2000) found that tumor-activated endothelium overexpressed specific transcripts as a result of qualitative differences in gene profiling compared with the normal endothelial cells of the tissue of origin. A total of 79 transcripts were differentially expressed: 46 were elevated at least 10-fold and 33 were expressed at substantially lower levels in tumor-associated endothelial cells. In glioblastoma, tumor endothelial cells carry the same genomic alteration as tumor cells (Ricci-Vitiani et al., 2010) and tumor cells directly transdifferentiate into $CD31^+$, $CD34^+$ endothelial cells, which may be involved in resistance toward anti-VEGF therapy (Ria et al., 2009; Soda et al., 2011). Ria et al. (2009) have identified genes differentially expressed in multiple myeloma (MM) endothelial cells as compared to monoclonal gammopathy of undetermined significance (MGUS) endothelial cells (Fig. 6.2). Deregulated genes were mostly involved in ECM formation and bone remodeling, cell adhesion, chemotaxis, angiogenesis, resistance to apoptosis, and cell cycle regulation. Tenascin-C positive human neuroblastoma cells transdifferentiate into tumor-derived endothelial cells which have been detected both in primary tumors and in tumors formed by human neuroblastoma cell lines in immunodeficient mice (Pezzolo et al., 2011). Tumor-derived endothelial cells targeted with cytotoxic human CD31 monoclonal antibody enhance tumor hypoxia, which in turn induces epithelial mesenchymal transition (EMT) and endothelial transdifferentiation of tumor cells (Pezzolo et al., 2014).

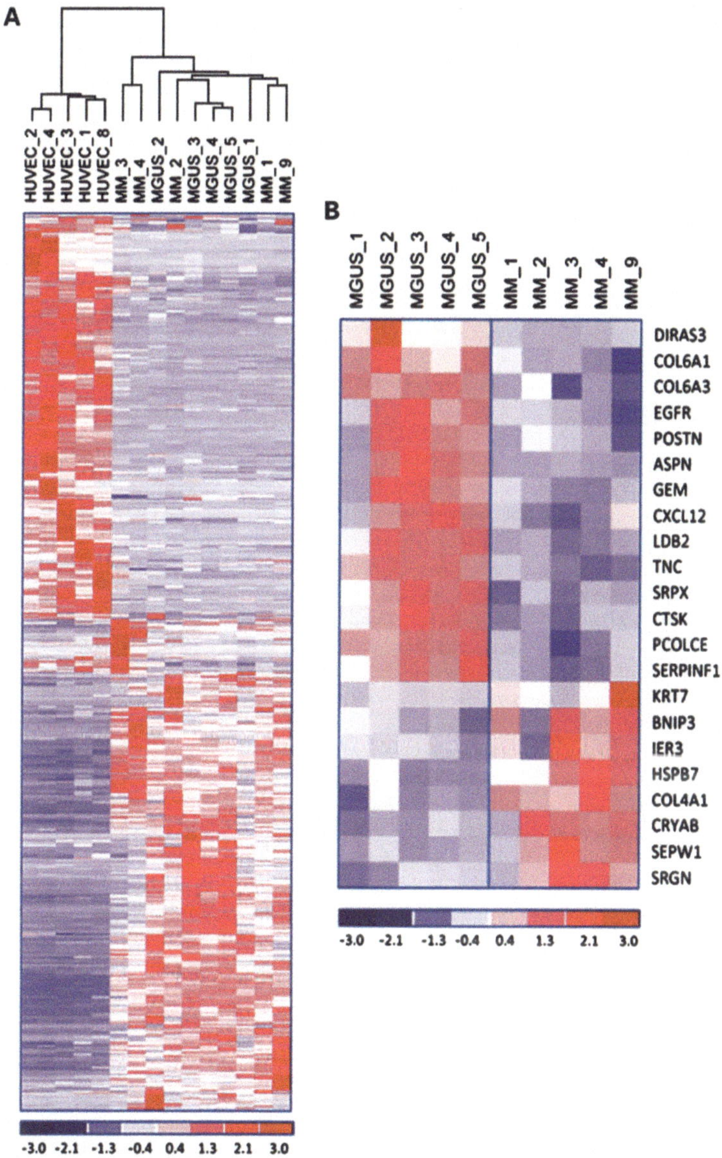

Figure 6.2 Unsupervised (A) and (B) supervised analysis of gene expression profiles from a data set composed of five monoclonal gammopathyies of undetermined significance endothelial cells, five multiple myeloma endothelial cells, and five HUVEC samples. In (B) identification of the 22 genes differentially expressed in five multiple myeloma endothelial cells versus five monoclonal gammopathy of undetermined significance endothelial cells. Color scale bar, the relative gene expression changes normalized by the SD; the color changes in each row is gene expression relative to the mean across the samples (with gene symbols). *(Reproduced form Ria, R., Todoerti, K., Berardi, S., Coluccia, A. M. L., De Luisi, A., Mattioli, M., et al (2009). Gene expression profiling of bone marrow endothelial cells in patients with multiple myeloma.* Clinical Cancer Research, 15*(17), 5369–5378. https://doi.org/10.1158/1078-0432.ccr-09-0040.)*

6.3 Note

1. According to the specific function of the host tissue or organ, capillary endothelial cells express further phenotypic differentiation and were classified accordingly: those endowed only with numerous caveolae have been termed continuous or nonfenestrated capillaries (as in the brain and heart), those bestowed with abundant fenestrae in addition to vesicles were called fenestrated capillaries (as in endocrine organs), and those that display large fenestrae without diaphragms were named discontinuous (as in bone marrow and liver).

References

Akiyama, K., Ohga, N., Hida, Y., Kawamoto, T., Sadamoto, Y., Ishikawa, S., et al. (2012). Tumor endothelial cells acquire drug resistance by MDR1 up-regulation via VEGF signaling in tumor microenvironment. *American Journal Of Pathology, 180*(3), 1283–1293. https://doi.org/10.1016/j.ajpath.2011.11.029.

Cantelmo, A. R., Conradi, L.-C., Brajic, A., Goveia, J., Kalucka, J., Pircher, A., et al. (2016). Inhibition of the glycolytic activator PFKFB3 in endothelium induces tumor vessel normalization, impairs metastasis, and improves chemotherapy. *Cancer Cell, 30*(6), 968–985. https://doi.org/10.1016/j.ccell.2016.10.006.

Croix, B. S. (2000). Genes expressed in human tumor endothelium. *Science, 289*(5482), 1197–1202. https://doi.org/10.1126/science.289.5482.1197.

Daldrup, H., Shames, D. M., Wendland, M., Okuhata, Y., Link, T. M., Rosenau, W., et al. (1998). Correlation of dynamic contrast-enhanced MR imaging with histologic tumor grade: Comparison of macromolecular and small-molecular contrast media. *American Journal of Roentgenology, 171*(4), 941–949. https://doi.org/10.2214/ajr.171.4.9762973.

Dvorak, H. F., Nagy, J. A., Dvorak, J. T., & Dvorak, A. M. (1988). Identification and characterization of the blood vessels of solid tumors that are leaky to circulating macromolecules. *American Journal of Pathology, 133*(1), 95–109.

Ferrara, N. (2002). Role of vascular endothelial growth factor in physiologic and pathologic angiogenesis: Therapeutic implications. *Seminars in Oncology, 29*(6), 10–14.

Giatromanolaki, A., Sivridis, E., Minopoulos, G., Polychronidis, A., Manolas, C., Simopoulos, C., et al. (2002). Differential assessment of vascular survival ability and tumor angiogenic activity in colorectal cancer. *Clinical Cancer Research, 8*(5), 1185–1191.

Goldmann, E. (1908). The growth of malignant disease in man and the lower animals, with special reference to the vascular system. *Proceedings of the Royal Society of Medicine, 1*, 1–13.

Hida, K., Maishi, N., Akiyama, K., Ohmura-Kakutani, H., Torii, C., Ohga, N., et al. (2017). Tumor endothelial cells with high aldehyde dehydrogenase activity show drug resistance. *Cancer Science, 108*(11), 2195–2203. https://doi.org/10.1111/cas.13388.

Karaman, S., & Detmar, M. (2014). Mechanisms of lymphatic metastasis. *Journal of Clinical Investigation, 124*(3), 922–928. https://doi.org/10.1172/JCI71606.

Kolonin, M., Pasqualini, R., & Arap, W. (2001). Molecular addresses in blood vessels as targets for therapy. *Current Opinion in Chemical Biology, 5*(3), 308–313. https://doi.org/10.1016/S1367-5931(00)00207-6.

Langenkamp, E., & Molema, G. (2009). Microvascular endothelial cell heterogeneity: General concepts and pharmacological consequences for anti-angiogenic therapy of cancer. *Cell and Tissue Research, 335*(1), 205–222. https://doi.org/10.1007/s00441-008-0642-4.

Magnussen, A., Kasman, I. M., Nerberg, S., Baluk, P., Murray, R., & McDonald, D. M. (2005). Rapid access of antibodies to α 5β 1 integrin overexpressed on the luminal surface of tumor blood vessels. *Cancer Research, 65*(7), 2712–2721. https://doi.org/10.1158/0008-5472.CAN-04-2691.

Mai, J., Virtue, A., Shen, J., Wang, H., & Yang, X. F. (2013). An evolving new paradigm: Endothelial cells - conditional innate immune cells. *Journal of Hematology and Oncology, 6*(1). https://doi.org/10.1186/1756-8722-6-61.

Masood, R., Gordon, E. M., Whitley, M. D., Wu, B. W., Cannon, P., Evans, L., et al. (2001). Retroviral vectors bearing IgG-binding motifs for antibody-mediated targeting of vascular endothelial growth factor receptors. *International Journal of Molecular Medicine, 8*(4), 335–343.

Matsuda, K., Ohga, N., Hida, Y., Muraki, C., Tsuchiya, K., Kurosu, T., et al. (2010). Isolated tumor endothelial cells maintain specific character during long-term culture. *Biochemical and Biophysical Research Communications, 394*(4), 947–954. https://doi.org/10.1016/j.bbrc.2010.03.089.

McDonald, D. M., & Choyke, P. L. (2003). Imaging of angiogenesis: From microscope to clinic. *Nature Medicine, 9*(6), 713–725. https://doi.org/10.1038/nm0603-713.

Pezzolo, A., Marimpietri, D., Raffaghello, L., Cocco, C., Pistorio, A., Gambini, C., et al. (2014). Failure of anti tumor-derived endothelial cell immunotherapy depends on augmentation of tumor hypoxia. *Oncotarget, 5*(21), 10368–10381. https://doi.org/10.18632/oncotarget.2015.

Pezzolo, A., Parodi, F., Marimpietri, D., Raffaghello, L., Cocco, C., Pistorio, A., et al. (2011). Oct-4^+/Tenascin C^+ neuroblastoma cells serve as progenitors of tumor-derived endothelial cells. *Cell Research, 21*(10), 1470–1486. https://doi.org/10.1038/cr.2011.38.

Ria, R., Todoerti, K., Berardi, S., Coluccia, A. M. L., De Luisi, A., Mattioli, M., et al. (2009). Gene expression profiling of bone marrow endothelial cells in patients with multiple myeloma. *Clinical Cancer Research, 15*(17), 5369–5378. https://doi.org/10.1158/1078-0432.ccr-09-0040.

Ribatti, D. (2005). The crucial role of vascular permeability factor/vascular endothelial growth factor in angiogenesis: A historical review. *British Journal of Haematology, 128*(3), 303–309. https://doi.org/10.1111/j.1365-2141.2004.05291.x.

Ribatti, D., Vacca, A., Nico, B., Fanelli, M., Roncali, L., & Dammacco, F. (1996). Angiogenesis spectrum in the stroma of B-cell non-Hodgkin's lymphomas. An immunohistochemical and ultrastructural study. *European Journal of Heaematology, 56*, 45–53.

Ricci-Vitiani, L., Pallini, R., Biffoni, M., Todaro, M., Invernici, G., Cenci, T., et al. (2010). Tumour vascularization via endothelial differentiation of glioblastoma stem-like cells. *Nature, 468*(7325), 824–828. https://doi.org/10.1038/nature09557.

Senger, D. R., Perruzzi, C. A., Feder, J., & Dvorak, H. F. (1986). A highly conserved vascular permeability factor secreted by a variety of human and rodent tumor cell lines. *Cancer Research, 46*(11), 5629–5632.

Soda, Y., Marumoto, T., Friedmann-Morvinski, D., Soda, M., Liu, F., Michiue, H., et al. (2011). Transdifferentiation of glioblastoma cells into vascular endothelial cells. *Proceedings of the National Academy of Sciences of the United States of America, 108*(11), 4274–4280. https://doi.org/10.1073/pnas.1016030108.

Stylianopoulos, T., Munn, L. L., & Jain, R. K. (2018). Reengineering the physical microenvironment of tumors to improve drug delivery and efficacy: From mathematical modeling to bench to bedside. *Trends in Cancer, 4*(4), 292–319. https://doi.org/10.1016/j.trecan.2018.02.005.

Thorpe, P. E. (2004). Vascular targeting agents as cancer therapeutics. *Clinical Cancer Research, 10*(2), 415–427. https://doi.org/10.1158/1078-0432.CCR-0642-03.

Vermeulen, P. B., Verhoeven, D., Hubens, G., Van Marck, E., Goovaerts, G., Huyghe, M., et al. (1995). Microvessel density, endothelial cell proliferation and tumour cell proliferation in human colorectal adenocarcinomas. *Annals of Oncology, 6*(1), 59–64.

CHAPTER 7

Tumor basement membrane

7.1 The structure of the basement membrane

The basement membrane is a highly specialized extracellular matrix sheet underlining epithelial or endothelial cells, consisting of collagen IV, laminins, entactin, and heparan sulfate proteoglycans, which are involved in the modulation of cell shape, gene expression, proliferation, migration, and apoptosis. By electron microscope observations, basement membranes are triple-layered structures consisting of a lamina lucida externa of variable width; an intermediate, 20−100 nm thick electron dense amorphous layer, the lamina densa or basal lamina, and the lamina lucida interna, 20−40 nm thick. In vivo studies in knockout mice for basement membrane genes (fibronectin, laminin, collagen IV, and perlecan) revealed significant cardiovascular dysfunctions (Astrof, Crowley, & Hynes, 2007; Pöschl et al., 2004; Thyboll et al., 2002).

Type IV collagen, the main protein component of all basement membranes, plays a crucial role in endothelial cell proliferation and cell behavior (Madri, 1997). Laminin is responsible for cessation of endothelial cell proliferation, pericyte recruitment, and vessel stabilization through Notch signaling activation (Estrach et al., 2011). Collagen IV and laminin are interconnected with proteoglycans such as nidogen and perlecan (Kalluri, 2003). The main functions of basement membrane are cell adhesion, diffusion barrier, and regulation of cell growth. First, it forms an adhesion interface between parenchymal cells and underlying extracellular matrix (ECM), the cells having adhesion mechanisms to anchor them to basement membrane. Second, the basement membrane acts as a permeability barrier with pore size depending on the charge and spatial arrangement of its glycosaminoglycans (GAGs). The basement membrane of blood vessels prevents large proteins leaking into the tissues. Third, the basement membrane controls cell organization and differentiation by the mutual interaction of cell surface receptors and molecules in the ECM.

Tumor Microenvironment Regulation of Tumor Expansion
ISBN 978-0-12-822803-6
https://doi.org/10.1016/B978-0-12-822803-6.00009-0

7.2 Tumor basement membrane

In the earliest stages of cancer, carcinomas in situ are separated from the tissue parenchyma by a basement membrane and to invade the tissue parenchyma tumor cells must acquire the ability to degrade the basement membrane. A functional vascular basement membrane is involved in controlling vessel permeability, intratumor edema, resistance to compression, bleeding, intravasation of tumor cells, and vessel perfusion. The vascular basement membrane in tumors has conspicuous abnormalities, including irregular thickness, multiple layers, and loose association with endothelial cells and pericytes, reminiscent of degenerating or regenerating blood vessels. Basement membrane also covers most endothelial sprouts. These features of the vascular basement membrane are consistent with the dynamic nature of endothelial cells and pericytes in tumors.

The abnormalities in tumor basement membrane can be partially explained by the disrupted endothelial cell–pericyte interactions (Baluk, Morikawa, Haskell, Mancuso, & McDonald, 2003). The basement membrane contains distinctive forms of fibronectin comprising the extra domain-B (ED-B), laminin, and type IV collagen with exposed cryptic sites (Santimaria et al., 2003), and proteoglycans that limit the expansion of normal microvessels to about 30 per cent (Chang et al., 2009). Fibronectin, laminin, collagen I, III, and IV play distinct and overlapping roles in angiogenesis. ECM is a source of angiogenic and antiangiogenic molecules. Three examples of antiangiogenic molecules are endostatin, which is a COOH- terminal fragment of collagen XVIII (Note 1), tumstatin and arresten, which are the noncollageneous-1 domain of the $\alpha3$ and the $\alpha1$ chains of type IV collagen, respectively (Colorado et al., 2000). Some of basement membrane structural proteins are broken down by enzymes to yield molecules with potent actions. Basement membrane degradation results from the increased expression and activation of pericyte cathepsin proteases, accompanied by decreased expression of cysteine protease inhibitors that limit cathepsin activity (Chang et al., 2009). Laminin, which consists of heterodimeric α, β, and γ chains, has been implicated in invasion and metastasis in various types of cancers (Akimoto et al., 2004). An anti-laminin peptide inhibits bone metastases of the A375 human melanoma cells and the MH-95 B-cell lymphoma cells (Michigami et al., 1998; Nakai, Mundy, Williams, Boyce, & Yoneda, 1992). Basement membrane is also a site of growth factor binding (Kalluri, 2003).

7.3 Note

1. Endostatin was originally found in conditioned media from a murine endothelial tumor cell line, hemangioendothelioma. Endostatin inhibits endothelial cell migration in vivo and in vitro and induces endothelial cell apoptosis, inhibits tumor growth, and impairs blood vessel maturation in wound healing. Finally, endostatin plays an important role in endothelial cell adhesion and cytoskeletal organization. Recombinant human endostatin (Endostar) has been widely applied for the clinical treatment of partial primary and metastatic solid tumors.

References

Akimoto, S., Nakanishi, Y., Sakamoto, M., Kanai, Y., & Hirohashi, S. (2004). Laminin 5 β3 and γ2 chains are frequently coexpressed in cancer cells. *Pathology International, 54*(9), 688–692. https://doi.org/10.1111/j.1440-1827.2004.01681.x.

Astrof, S., Crowley, D., & Hynes, R. O. (2007). Multiple cardiovascular defects caused by the absence of alternatively spliced segments of fibronectin. *Developmental Biology, 311*(1), 11–24. https://doi.org/10.1016/j.ydbio.2007.07.005.

Baluk, P., Morikawa, S., Haskell, A., Mancuso, M., & McDonald, D. M. (2003). Abnormalities of basement membrane on blood vessels and endothelial sprouts in tumors. *American Journal of Pathology, 163*(5), 1801–1815. https://doi.org/10.1016/S0002-9440(10)63540-7.

Chang, S.-H., Kanasaki, K., Gocheva, V., Blum, G., Harper, J., Moses, M. A., et al. (2009). VEGF-A induces angiogenesis by perturbing the cathepsin-cysteine protease inhibitor balance in venules, causing basement membrane degradation and mother vessel formation. *Cancer Research, 69*(10), 4537–4544. https://doi.org/10.1158/0008-5472.CAN-08-4539.

Colorado, P. C., Torre, A., Kamphaus, G., Maeshima, Y., Hopfer, H., Takahashi, K., et al. (2000). Anti-angiogenic cues from vascular basement membrane collagen. *Cancer Research, 60*(9), 2520–2526.

Estrach, S., Cailleteau, L., Franco, C. A., Gerhardt, H., Stefani, C., Lemichez, E., et al. (2011). Laminin-binding integrins induce Dll4 expression and Notch signaling in endothelial cells. *Circulation Research, 109*(2), 172–182. https://doi.org/10.1161/circresaha.111.240622.

Kalluri, R. (2003). Basement membranes: Structure, assembly and role in tumour angiogenesis. *Nature Reviews Cancer, 3*(6), 422–433. https://doi.org/10.1038/nrc1094.

Madri, J. A. (1997). Extracellular matrix modulation of vascular cell behaviour. *Transplant Immunology, 5*(3), 179–183. https://doi.org/10.1016/S0966-3274(97)80035-4. Elsevier B.V.

Michigami, T., Nomizu, M., Yamada, Y., Dunstan, C., Williams, P. J., Munday, G. R., et al. (1998). Growth and dissemination of a newly-established murine B-cell lymphoma cell line is inhibited by multimeric YIGSR peptide. *Clinical and Experimental Metastasis, 16*(7), 645–654. https://doi.org/10.1023/A:1006502528268.

Nakai, M., Mundy, G. R., Williams, P. J., Boyce, B., & Yoneda, T. (1992). A synthetic antagonist to laminin inhibits the formation of osteolytic metastases by human melanoma cells in nude mice. *Cancer Research, 52*(19), 5395–5399. https://www.ncbi.nlm.nih.gov/pubmed/1394144.

Pöschl, E., Schlötzer-Schrehardt, U., Brachvogel, B., Saito, K., Ninomiya, Y., & Mayer, U. (2004). Collagen IV is essential for basement membrane stability but dispensable for initiation of its assembly during early development. *Development, 131*(7), 1619–1628. https://doi.org/10.1242/dev.01037.

Santimaria, M., Moscatelli, G., Viale, G. L., Giovannoni, L., Neri, G., Viti, F., et al. (2003). Immunoscintigraphic detection of the ED-B domain of fibronectin, a marker of angiogenesis, in patients with cancer. *Clinical Cancer Research, 9*(2), 571–579. https://www.ncbi.nlm.nih.gov/pubmed/12576420.

Thyboll, J., Kortesmaa, J., Cao, R., Soininen, R., Wang, L., Iivanainen, A., et al. (2002). Deletion of the laminin α4 chain leads to impaired microvessel maturation. *Molecular and Cellular Biology, 22*(4), 1194–1202. https://doi.org/10.1128/MCB.22.4.1194-1202.2002.

CHAPTER 8

Tumor pericytes

8.1 Morphological and functional characteristics of pericytes

During the initial phase of angiogenesis, activated pericytes in parent vessels bulge, shorten their processes, and increase their volume and an intense cell proliferation occurs, where they project into the perivascular spaces, where their basement membrane is disrupted and fragmented and appears detached from the vessel wall (Diaz-Flores, Gutierrez, & Varela, 1992). Although initially endothelial cell sprouts may form without pericyte involvement, pericytes are among the first cells to invade newly vascularized tissues and locate at the growing front of the endothelial sprouts by determining the location of sprout formation and by guiding newly formed vessels (Fig. 8.1) (Nehls, Denzer, & Drenckhahn, 1992).

Even though pericytes were described for the first time by Eberth already in 1871 and Rouget in 1873, most of the progress in studying pericytes has been made later with the introduction of more advanced microscopy techniques as well as the generation of transgenic animal models which enable studying pericyte roles under in vivo conditions. A universally accepted definition of pericytes does not exist. Therefore, when identifying a pericyte, several criteria need to be taken into account, including their morphology and location, as well as specific transcriptome/proteome expressions. No single universal pericyte marker has been identified. Different subpopulations of pericytes have been identified based on marker expression and as a progenitor cells for different subsets. However, the function of individual pericyte populations in tumor biology is poorly explored. Pericytes have been characterized by the expression of alpha-smooth actin (α-SMA), platelet derived growth factor receptor beta (PDGFRβ), CD146, and nerve/glial antigen 2 (NG2) proteoglycan (Ribatti, Nico, & Crivellato, 2011). Pericytes are usually found as a single cell or a discontinuous single cell layer around arterioles, capillaries, and postcapillary venules.

Tumor Microenvironment Regulation of Tumor Expansion
ISBN 978-0-12-822803-6
https://doi.org/10.1016/B978-0-12-822803-6.00012-0

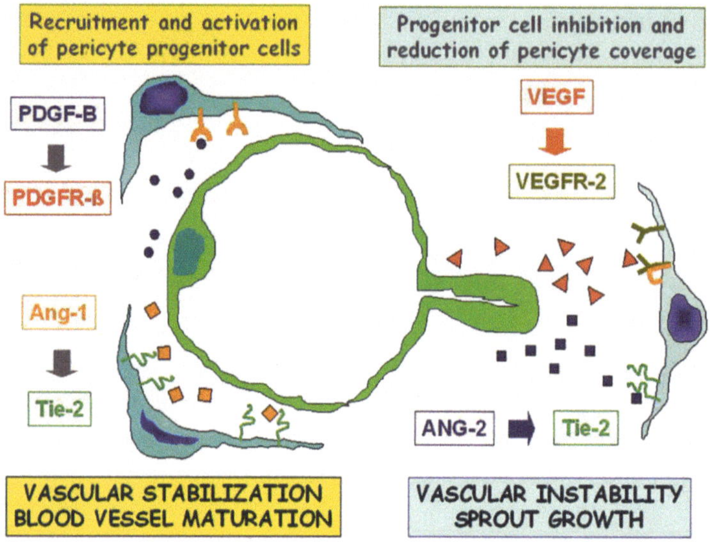

Figure 8.1 Schematic drawing that illustrates the paracrine interactions occurring between pericyte precursor cells and endothelial cells in PDGF-mediated angiogenesis. Endothelial cells secrete PDGF-B that causes pericyte precursor cell proliferation and migration through activation of PDGFR-B receptors. Pericytes surround and cover early endothelial tubes. By contrast, endothelial cells in vascular sprouts release VEGF, which in turn mediates suppression of PDGFR-β signaling through the induction of VEGFR-2/PDGFR-β complexes. This pathway abrogates pericyte coverage of endothelial sprouts leading to vascular instability and regression. *(Reproduced from Ribatti, D., Nico, B., & Crivellato, E. (2011). The role of pericytes in angiogenesis.* International Journal of Developmental Biology, *55(3), 261–268. https://doi.org/10.1387/ijdb.103167dr.)*

Pericytes are recruited by PDGF secreted by activated endothelium, and PDGF function critically relies on its interaction with extracellular matrix (ECM). Removal of matrix-binding domains leads to pericyte detachment from nascent vessels and severe vascular dysfunction (Lindblom, 2003). Transforming growth factor beta 1 (TGFβ1) is synthesized in a latent form and its activation requires cell to cell contact between endothelial and mural cells through proteolytic cleavage (Sato & Rifkin, 1989). Angiopoietin-1 (Ang-1) facilitates recruitment and association of pericytes with newly formed vascular channels, acts as a survival signal for endothelial cells, and inhibits vascular endothelial growth factor (VEGF)-induced vascular leakage (Fig. 8.2) (Fagioni & Christifori, 2013). On the other hand, Ang-2 promotes the dissociation of pericytes and endothelial cells, and endothelial cell apoptosis (Fig. 8.2) (Fagioni & Christifori, 2013). Pericytes interact with

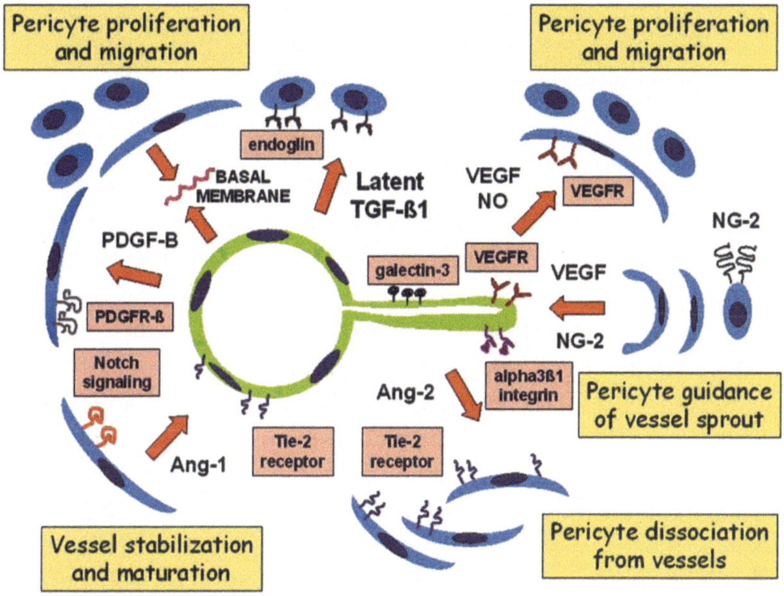

Figure 8.2 Signaling pathways operating in endothelial cell/pericyte paracrine cross-talk. Pericytes are involved in endothelial cell stimulation and guidance as well as endothelial stabilization and maturation. Vessel sprouts (right) cause destabilization of pericyte investment through Ang-2/Tie-2 signaling. Pericytes provide guidance for endothelial movement and tube formation through secretion of VEGF and soluble NG-2. Spreading endothelial cells, in turn, stimulate pericyte precursor cell proliferation and migration by releasing VEGF and NO. Vessel stabilization (left) occurs by pericyte investment and close interaction with endothelial cells. Mature endothelial cells secrete PDGF-B, which promotes proliferation and migration of pericyte precursor cells through activation of PDGFR-B receptors expressed on the surface of pericyte progenitors. This mechanism leads to pericyte coverage of early endothelial tubes. Vessel maturation further develops through Ang-1- and Notch-mediated signaling. Pericytes stabilize and reinforce the endothelial tube contributing to secretion of basement membrane. *(Reproduced from Ribatti, D., Nico, B., & Crivellato, E. (2011). The role of pericytes in angiogenesis.* International Journal of Developmental Biology, *55(3), 261–268. https://doi.org/10.1387/ijdb.103167dr.)*

the endothelium through the stromal cell derived factor-1 (SDF-1)/C-X-C chemokine receptor type 4 (CXCR4) axis, which is responsible for their recruitment and migration along the vessels (Song et al., 2009). In addition, the interaction between endothelial Jagged-1 and pericyte Nothc3 is responsible for vessel stabilization (Hua, Simone, & Brenda, 2009). Birbrair et al. (2013) identified two pericyte subpopulations from large blood vessels and capillaries, named type-1 and type-2 pericytes. In vitro and in vivo assays revealed that type-2 (Nestin-GFP$^+$/NG2-DsRed$^+$), but not type-1 (Nestin-GFP$^-$/NG2$^-$DsRed$^-$) pericytes are recruited during tumor angiogenesis (Ribeiro & Okamoto, 2015).

8.2 Tumor pericytes

In tumors, NG2 is a marker of a mature population of pericytes (Song, Ewald, Stallcup, Werb, & Bergers, 2005). Desmin positivity represents a more mature population of pericytes (Song et al., 2005). Desmin-positive pericytes appear more closely associated with vessels, whereas αSMA-positive and PDGFR-β^+ are partially detached (Morikawa et al., 2002). Pericytes may cover 73–92 percent of endothelial sprouts in several murine tumor types, and are located near blood vessels at the growing front of tumors, where angiogenesis is most active and show morphological abnormalities (Morikawa et al., 2002; Schlingemann, Rietveld, De Waal, Ferrone, & Ruiter, 1990; Wesseling et al., 1995). Breast and colon tumors recruit significantly more pericytes than gliomas or renal cell carcinoma (Eberhard et al., 2000). Pericyte deficiency could be partly responsible for vessel abnormalities in tumor blood vessels (Gerhardt & Semb, 2008), and partial dissociation of pericytes (Hashizume et al., 2000) contribute to increased tumor vascular permeability. Severe reduction or lack of pericyte coverage may disrupt the integrity of the vasculature, enabling tumor cells to transit into the circulatory system, thereby facilitating metastasis. Genetic disruption of pericyte coverage elicited increased metastasis in the Rip1-Tag2 pancreatic islet tumor model (Note 1) (Xian et al., 2006). Moreover, the increased interstitial pressure compress other tumor vessels, triggering tumor metastasis through a hypoxia-induced epithelial mesenchymal transition (EMT) mechanism (Cooke et al., 2012).

Tumor pericytes are loosely associated with endothelial cells, have abnormal shape, paradoxically extend cytoplasmic processes away from the vessel wall, and have extra layers of loosely fitting basement membrane (Eberhard et al., 2000). This abnormal pericyte–endothelial cell association contributes to increased vascular permeability (Hashizume et al., 2000). Their abnormalities are consistent with alterations in PDGF signaling pathways. Mice genetically deficient in PDGFB or its receptors have blood vessels with loose pericyte attachment, irregular vessel caliber, luminal projections on endothelial cell, and hemorrhage. Similar abnormalities occur in many tumor vessels. PDGFB expressed by tumor cells increases pericyte recruitment in several in vivo models (Guo et al., 2003).

Alternatively, genetic abolition of PDGFB-R expressed by embryonic pericytes decreased their recruitment (Abramsson, Lindblom, & Betsholtz, 2003). In Lewis lung carcinoma implanted in mice, inhibition of endothelial differentiation gene-1 (EDG-1) expression in endothelial cell strongly reduced pericyte coverage (Chae, Paik, Furneaux, & Hla, 2004). In a human glioma model developed in rat, Ang-1 led to enhanced pericyte recruitment and increased tumor growth, presumably by favoring angiogenesis (Machein et al., 2004). Alternatively, in a colon cancer model, overexpression of Ang-1 led to smaller tumors with fewer blood vessels and greater pericyte coverage, decreased vascular permeability, and reduced hepatic metastasis (Ahmad et al., 2001).

The pericyte coverage of the endothelium is variable, depending on the type of tumor. Islet carcinomas have dense pericyte coverage, while glioblastomas have a reduced number of pericytes (Bergers & Song, 2005). In a human neuroblastoma xenotransplanted model, pericyte coverage is halved in tumors grafted on matrix metalloproteinase-9 (MMP-9)-deficient mice and transplantation with MMP-9-expressing bone marrow cells restores the formation of mature vessels (Chantrain et al., 2004). In addition, overexpression of tissue inhibitor of MMP-3 (TIMP-3) results in decreased pericyte recruitment in neuroblastoma and melanoma models (Spurbeck, Ng, Strom, Vanin, & Davidoff, 2002). The observation that pericytes express MMPs in many human tumors in vivo (Nielsen et al., 1997) suggests that pericyte invasion requires the proteolytic degradation of ECM by proteases, including MMPs. VEGF inhibition eliminates tumor vessels without removing pericytes (Morikawa et al., 2002). Antiangiogenic treatment directed against endothelial cells using VEGF inhibitors induces the regression of tumor vessels and decreased tumor size (Baluk, Hashizume, & McDonald, 2005), leading to vessel normalization, characterized by increased pericyte coverage, tumor perfusion, and chemotherapeutic sensitivity (Jain, 2005). Moreover, removal of VEGF inhibition causes tumor regrowth due to the fact that pericytes provide a scaffold for the rapid regrowth of tumor vessels (Mancuso et al., 2006). Pericytes have been indicated as putative targets in the pharmacological therapy of tumors by using the synergistic effect of antiendothelial and antipericytic molecules.

Removal of pericyte coverage leads to exposed tumor vessels, which may explain the enhanced effect of combining inhibitors that target both tumor vessels and pericytes. Bergers, Song, Meyer-Morse, Bergsland, & Hanahan (2003) showed that combined treatment or pre-treatment with anti-PDGF-B/PDGFBR-β reducing pericyte coverage increases the success of anti-VEGF treatment in the mouse RIP1-TAG2 model. However, extensive regression of endothelial cells was not observed in tumors after inhibition of PDGFR-β signaling (Abramsson et al., 2003). STI571 (Gleevec, Imatinib), which targets PDGFRs and other receptor tyrosine kinases, did not reduce vascular density when given alone but did augment the effects of VEGF inhibitors (Bergers et al., 2003). After treatment of RIP1-TAG-2 tumors and Lewis lung carcinomas with AG-013737 or VEGF-Trap, surviving pericytes may become more tightly associated with endothelial cells or have no apparent association with tumor vessels (Inai et al., 2004). Treatment of RIP1-TAG2 tumors with anti-PDGFR-β antibody for 3 weeks reduces pericytes, increases endothelial cell apoptosis, but does not seem to reduce tumor vascular density (Song et al., 2005). Similarly, the receptor tyrosine kinase inhibitor SU6668, which also affects PDGFR-β signaling, detaches and reduces pericytes in RIP1-TAG2 and xenotransplanted tumors, thereby restricting tumor growth (Reinmuth et al., 2001; Shaheen et al., 2001). Sennino et al. (2007) demonstrated that treatment with a novel DNA oligonucleotide aptamer (AX102) that selectively binds PDGF-B led to progressive reduction of pericytes in Lewis lung carcinomas. Murphy et al. (2010) generated a series of selective type II inhibitors of PDGFR-β and B-RAF targets for pericyte recruitment and endothelial survival, respectively, and demonstrated that dual inhibition of both PDGFR-β and B-RAF exerted synergistic antiangiogenic activity in both zebrafish and murine models of angiogenesis. A high prevalence of abluminal pericytes limits tumor metastasis (Xian et al., 2006), wheras deficient pericyte coverage can be associated with increased metastatic potential (Yonenaga et al., 2005). Pericytes have also a direct role in tumor immune escape by inhibiting CD4 T cell activation and by promoting T cell anergy (Bose et al., 2013).

8.3 Note

1. The Rip-Tag2 mouse model of pancreatic islet cell carcinoma was developed by Doug Hanahan in 1985 and offers a model of spontaneous multistep tumorigenesis. Transgenic mice express the oncogene SV40 T antigen (Tag) under the control of the rat insulin gene promoter (Rip) and display different stages of tumor progression: onset of hyperproliferation, induction of angiogenesis and formation of solid tumors. Tumor development proceeds by stages during which about half the 400 islets become hyperproliferate, while a subset (about 25 percent)

subsequently acquire the ability to switch to angiogenesis.3. Some 15–20 percent of these angiogenic islets develop into benign tumors, encapsulated lesions, and invasive carcinomas.

References

Abramsson, A., Lindblom, P., & Betsholtz, C. (2003). Endothelial and nonendothelial sources of PDGF-B regulate pericyte recruitment and influence vascular pattern formation in tumors. *Journal of Clinical Investigation, 112*(8), 1142–1151. https://doi.org/10.1172/JCI200318549.

Ahmad, S. A., Liu, W., Jung, Y. D., Fan, F., Wilson, M., Reinmuth, N., et al. (2001). The effects of angiopoietin-1 and -2 on tumor growth and angiogenesis in human colon cancer. *Cancer Research, 61*(4), 1255–1259.

Baluk, P., Hashizume, H., & McDonald, D. M. (2005). Cellular abnormalities of blood vessels as targets in cancer. *Current Opinion in Genetics and Development, 15*(1), 102–111. https://doi.org/10.1016/j.gde.2004.12.005.

Bergers, G., & Song, S. (2005). The role of pericytes in blood-vessel formation and maintenance. *Neuro-Oncology, 7*(4), 452–464. https://doi.org/10.1215/s1152851705000232.

Bergers, G., Song, S., Meyer-Morse, N., Bergsland, E., & Hanahan, D. (2003). Benefits of targeting both pericytes and endothelial cells in the tumor vasculature with kinase inhibitors. *Journal of Clinical Investigation, 111*(9), 1287–1295. https://doi.org/10.1172/JCI200317929.

Birbrair, A., Zhang, T., Wang, Z.-M., Messi, M. L., Enikolopov, G. N., Mintz, A., et al. (2013). Role of pericytes in skeletal muscle regeneration and fat accumulation. *Stem Cells and Development, 22*(16), 2298–2314. https://doi.org/10.1089/scd.2012.0647.

Bose, A., Barik, S., Banerjee, S., Ghosh, T., Mallick, A., Bhattacharyya Majumdar, S., et al. (2013). Tumor-derived vascular pericytes anergize Th cells. *The Journal of Immunology, 191*(2), 971–981. https://doi.org/10.4049/jimmunol.1300280.

Chae, S. S., Paik, J. H., Furneaux, H., & Hla, T. (2004). Requirement for sphingosine 1-phosphate receptor-1 in tumor angiogenesis demonstrated by in vivo RNA interference. *Journal of Clinical Investigation, 114*(8), 1082–1089. https://doi.org/10.1172/JCI200422716.

Chantrain, C. F., Shimada, H., Jodele, S., Groshen, S., Ye, W., Shalinsky, D. R., et al. (2004). Stromal matrix metalloproteinase-9 regulates the vascular architecture in neuroblastoma by promoting pericyte recruitment. *Cancer Research, 64*(5), 1675–1686. https://doi.org/10.1158/0008-5472.CAN-03-0160.

Cooke, V. G., LeBleu, V. S., Keskin, D., Khan, Z., O'Connell, J. T., Teng, Y., et al. (2012). Pericyte depletion results in hypoxia-associated epithelial-to-mesenchymal transition and metastasis mediated by met signaling pathway. *Cancer Cell, 21*(1), 66–81. https://doi.org/10.1016/j.ccr.2011.11.024.

Diaz-Flores, L., Gutierrez, R., & Varela, H. (1992). Behavior of postcapillary venule pericytes during postnatal angiogenesis. *Journal of Morphology, 213*(1), 33–45. https://doi.org/10.1002/jmor.1052130105.

Eberhard, A., Kahlert, S., Goede, V., Hemmerlein, B., Plate, K. H., & Augustin, H. G. (2000). Heterogeneity of angiogenesis and blood vessel maturation in human tumors: Implications for antiangiogenic tumor therapies. *Cancer Research, 60*(5), 1388–1393.

Fagioni, E., & Christifori, G. (2013). Angiopoietin in angiogenesis. *Cancer Letters, 328*, 18–26.

Gerhardt, H., & Semb, H. (2008). Pericytes: gatekeepers in tumour cell metastasis? *Journal of Molecular Medicine, 86*(2), 135–144. https://doi.org/10.1007/s00109-007-0258-2.

Guo, P., Hu, B., Gu, W., Xu, L., Wang, D., Huang, H. J. S., et al. (2003). Platelet-derived growth factor-B enhances glioma angiogenesis by stimulating vascular endothelial growth factor expression in tumor endothelia and by promoting pericyte recruitment. *American Journal of Pathology, 162*(4), 1083–1093. https://doi.org/10.1016/S0002-9440(10)63905-3.

Hashizume, H., Baluk, P., Morikawa, S., McLean, J. W., Thurston, G., Roberge, S., et al. (2000). Openings between defective endothelial cells explain tumor vessel leakiness. *American Journal of Pathology, 156*(4), 1363–1380. https://doi.org/10.1016/S0002-9440(10)65006-7.

Hua, L., Simone, K., & Brenda, L. (2009). NOTCH3 expression is induced in mural cells through an autoregulatory loop that requires endothelial-expressed JAGGED1. *Circulation Research*, 466–475. https://doi.org/10.1161/circresaha.108.184846.

Inai, T., Mancuso, M., Hashizume, H., Baffert, F., Haskell, A., Baluk, P., et al. (2004). Inhibition of vascular endothelial growth factor (VEGF) signaling in cancer causes loss of endothelial fenestrations, regression of tumor vessels, and appearance of basement membrane ghosts. *American Journal of Pathology, 165*(1), 35–52. https://doi.org/10.1016/s0002-9440(10)63273-7.

Jain, R. K. (2005). Normalization of tumor vasculature: An emerging concept in anti-angiogenic therapy. *Science, 307*(5706), 58–62. https://doi.org/10.1126/science.1104819.

Lindblom, P. (2003). Endothelial PDGF-B retention is required for proper investment of pericytes in the microvessel wall. *Genes and Development, 17*(15), 1835–1840. https://doi.org/10.1101/gad.266803.

Machein, M. R., Knedla, A., Knoth, R., Wagner, S., Neuschl, E., & Plate, K. H. (2004). Angiopoietin-1 promotes tumor angiogenesis in a rat glioma model. *American Journal of Pathology, 165*(5), 1557–1570. https://doi.org/10.1016/S0002-9440(10)63413-X.

Mancuso, M. R., Davis, R., Norberg, S. M., O'Brien, S., Sennino, B., Nakahara, T., et al. (2006). Rapid vascular regrowth in tumors after reversal of VEGF inhibition. *Journal of Clinical Investigation, 116*(10), 2610–2621. https://doi.org/10.1172/jci24612.

Morikawa, S., Baluk, P., Kaidoh, T., Haskell, A., Jain, R. K., & McDonald, D. M. (2002). Abnormalities in pericytes on blood vessels and endothelial sprouts in tumors. *American Journal of Pathology, 160*(3), 985–1000. https://doi.org/10.1016/S0002-9440(10)64920-6.

Murphy, E. A., Shields, D. J., Stoletov, K., Dneprovskaia, E., McElroy, M., Greenberg, J. I., et al. (2010). Disruption of angiogenesis and tumor growth with an orally active drug that stabilizes the inactive state of PDGFR/B-RAF. *Proceedings of the National Academy of Sciences, 107*(9), 4299–4304. https://doi.org/10.1073/pnas.0909299107.

Nehls, V., Denzer, K., & Drenckhahn, D. (1992). Pericyte involvement in capillary sprouting during angiogenesis in situ. *Cell and Tissue Research, 270*(3), 469–474. https://doi.org/10.1007/BF00645048.

Nielsen, B. S., Sehested, M., Kjeldsen, L., Borregaard, N., Rygaard, J., & Danø, K. (1997). Expression of matrix metalloprotease-9 in vascular pericytes in human breast cancer. *Laboratory Investigation, 77*(4), 345–355.

Reinmuth, N., Liu, W., Jung, Y. D., Ahmad, S. A., Shaheen, R. M., Fan, F., et al. (2001). Induction of VEGF in perivascular cells defines a potential paracrine mechanism for endothelial cell survival. *The FASEB Journal, 15*(7), 1239–1241. Official Publication of the Federation of American Societies for Experimental Biology.

Ribatti, D., Nico, B., & Crivellato, E. (2011). The role of pericytes in angiogenesis. *International Journal of Developmental Biology, 55*(3), 261–268. https://doi.org/10.1387/ijdb.103167dr.

Ribeiro, A. L., & Okamoto, O. K. (2015). Combined effects of pericytes in the tumor microenvironment. *Stem Cells International, 2015*. https://doi.org/10.1155/2015/868475.

Sato, Y., & Rifkin, D. B. (1989). Inhibition of endothelial cell movement by pericytes and smooth muscle cells: Activation of a latent transforming growth factor-β1-like molecule by plasmin during co-culture. *Journal of Cell Biology, 109*(1), 309–315. https://doi.org/10.1083/jcb.109.1.309.

Schlingemann, R. O., Rietveld, F. J. R., De Waal, R. M. W., Ferrone, S., & Ruiter, D. J. (1990). Expression of the high molecular weight melanoma-associated antigen by pericytes during angiogenesis in tumors and in healing wounds. *American Journal of Pathology, 136*(6), 1393–1405.

Sennino, B., Falcón, B. L., McCauley, D., Le, T., McCauley, T., Kurz, J. C., et al. (2007). Sequential loss of tumor vessel pericytes and endothelial cells after inhibition of platelet-derived growth factor B by selective aptamer AX102. *Cancer Research, 67*(15), 7358–7367. https://doi.org/10.1158/0008-5472.CAN-07-0293.

Shaheen, R. M., Tseng, W. W., Davis, D. W., Liu, W., Reinmuth, N., Vellagas, R., et al. (2001). Tyrosine kinase inhibition of multiple angiogenic growth factor receptors improves survival in mice bearing colon cancer liver metastases by inhibition of endothelial cell survival mechanisms. *Cancer Research, 61*(4), 1464–1468. https://www.ncbi.nlm.nih.gov/pubmed/11245452.

Song, S., Ewald, A. J., Stallcup, W., Werb, Z., & Bergers, G. (2005). PDGFRβ^{+} perivascular progenitor cells in tumours regulate pericyte differentiation and vascular survival. *Nature Cell Biology,* 7(9), 870–879. https://doi.org/10.1038/ncb1288.

Song, N., Huang, Y., Shi, H., Yuan, S., Ding, Y., Song, X., et al. (2009). Overexpression of platelet-derived growth factor-BB increases tumor pericyte content via stromal-derived factor-1α/CXCR4 axis. *Cancer Research, 69*(15), 6057–6064. https://doi.org/10.1158/0008-5472.CAN-08-2007.

Spurbeck, W. W., Ng, C. Y. C., Strom, T. S., Vanin, E. F., & Davidoff, A. M. (2002). Enforced expression of tissue inhibitor of matrix metalloproteinase-3 affects functional capillary morphogenesis and inhibits tumor growth in a murine tumor model. *Blood, 100*(9), 3361–3368. https://doi.org/10.1182/blood.V100.9.3361.

Wesseling, P., Schlingemann, R. O., Rietveld, F. J. R., Link, M., Burger, P. C., & Ruiter, D. J. (1995). Early and extensive contribution of pericytes/vascular smooth muscle cells to microvascular proliferation in glioblastoma multiforme: An immuno-light and immuno-electron microscopic study. *Journal of Neuropathology and Experimental Neurology, 54*(3), 304–310. https://doi.org/10.1097/00005072-199505000-00003.

Xian, X., Håkansson, J., Ståhlberg, A., Lindblom, P., Betsholtz, C., Gerhardt, H., et al. (2006). Pericytes limit tumor cell metastasis. *Journal of Clinical Investigation, 116*(3), 642–651. https://doi.org/10.1172/JCI25705.

Yonenaga, Y., Mori, A., Onodera, H., Yasuda, S., Oe, H., Fujimoto, A., et al. (2005). Absence of smooth muscle actin-positive pericyte coverage of tumor vessels correlates with hematogenous metastasis and prognosis of colorectal cancer patients. *Oncology, 69*(2), 159–166. https://doi.org/10.1159/000087840.

CHAPTER 9

Inflammatory cells in tumor microenvironment

9.1 The link between chronic inflammation and tumorigenesis

Tumor cells undergo a Darwinian selection and are able to survive and enter into an equilibrium state in which the innate and adaptive immune system control the tumor growth (Dunn, Bruce, Ikeda, Old, & Schreiber, 2002; Schreiber, Old, & Smyth, 2011). Some tumor cells acquire mutations, chromosome amplifications and deletions, and epigenetic modifications, resulting in gene silencing or synthesis of abnormal proteins. Overall, these events allow tumor cells to escape the control of the immune system, proliferate, and give rise to a clinically detectable tumor. The historical observations made by Paul Ehrlich (Fig. 9.1) that tumor cells are recognized and eliminated by the immune cells evolved in the theory of innate immune surveillance, proposed by Sir Macfarlane Burnet (Fig. 9.2) in 1970 (Note 1). In this context, the immune system spontaneously identifies and eliminates cancer cells. According to the concept of immunoediting, tumor cells that are vulnerable to attack by the immune system are cleared, whereas cells that have a capacity to circumvent immunosurveillance can survive. Immunodeficient mice are more prone to the development of spontaneous tumors, and carcinogen-induced sarcoma developed better in a host devoid of immunity (Shankaran et al., 2001). Patients who are immunodeficient, and particularly those who have deficits in T cell immunity, are at increased risk of cancers, especially those caused by oncogenic viruses.

The link between chronic inflammation and tumorigenesis was first proposed by Rudolf Virchow in 1863 after the observation that infiltrating leukocytes are a hallmark of tumors and first established a causative connection between the lymphoreticular infiltrate at sites of chronic

Tumor Microenvironment Regulation of Tumor Expansion
ISBN 978-0-12-822803-6
https://doi.org/10.1016/B978-0-12-822803-6.00003-X

Figure 9.1 A portrait of Paul Ehrlich. *(From Wikipedia).*

Figure 9.2 A portrait of Sir Frank Macfarlane Burnet. *(From Wikipedia).*

Table 9.1 Chronic inflammatory states and cancer.

Pathologic condition and associated tumor(s)
Asbestosis, silicosis (mesothelioma, lung carcinoma)
Inflammatory bowel disease (colorectal carcinoma)
Pancreatitis (pancreatic carcinoma)
Chronic cholecystitis (gallbladder cancer)
Reflux esophagitis, Barrett's esophagus (esophageal carcinoma)
Sjögren syndrome, Hashimoto's thyroiditis (MALT lymphoma)
Opisthorchis, cholangitis (cholangiocarcinoma, colon carcinoma)
Gastritis/ulcers (gastric adenocarcinoma, MALT lymphoma)
Hepatitis (hepatocellular carcinoma)
Osteomyelitis (carcinoma in draining sinuses)
Chronic cervicitis (cervical carcinoma)
Chronic cystitis (Baldder carcinoma)

inflammation and the development of cancer (Table 9.1) (Balkwill, Charles, & Mantovani, 2005). Surgeons have long described the tendency of tumors to recur in healing resection margin, and it has been reported that wound healing environment provides an opportunistic matrix for tumor growth (Abramovitch, Marikovsky, Meir, & Neeman, 1998; Wong & Reiter, 2011). The association between inflammation and cancer is best documented in the digestive tract. For example, chronic ulcerative colitis and Crohn's disease increase the risk of colon cancer, acid reflux is associated with esophageal cancer, hepatitis C infection and alcohol consumption increase the risk of hepatocellular carcinoma, and *Helicobacter pylori* infection increases the risk of gastric and colon cancer. Finally, inflammation is also implicated in obesity-related cancer.

Tumors were described by Harold Dvorak (Fig. 9.3) as wounds that never heal (Dvorak, 1986). As Dvorak pointed out in 2015: "The novelty of my 1986 article was based on two, at the time, recent findings, namely the discovery of vascular permeability factor (VPF) as tumor product and the recognition that the chronic vascular hyperpermeability induced by VPF likely accounted for the fibrin deposited in solid tumors and in early stages of wound healing. Together, these findings anticipated that tumors and wound healing could be linked together in a fundamental way at the molecular level" (Dvorak, 2015). Wound healing is usually self-limiting, whereas tumors secrete vascular endothelial growth factor (VEGF) that can lead to persistent extravasation of fibrin and fibronectin and continuous generation of extracellular matrix (ECM). Wounds in rodent skin, like tumors, secrete VPF/VEGF: within 24 h of wounding VPF/VEGF mRNA expression increases in epidermal keratinocytes at the wound edge

Figure 9.3 A portrait of Harold Dvorak.

(Brown et al., 1992). VPF/VEGF overexpression reaches a peak at 2–3 days and persists at an elevated level for about 1 week, the time required for granulation tissue to form and migrating keratinocytes to cover the wound defect. In contrast to tumors, VPF/VEGF expression was downregulated as healing progressed and, parallel with the decreased expression of VPF/VEGF, vascular permeability returned to normal.

VPF/VEGF is among the most potent vascular permeabilizing agents known, with a potency some 50,000 times that of histamine and it is effective at concentrations well below 1 nM in the Miles assay which provides a sensitive measure of increased microvascular permeability to plasma proteins (Fig. 9.4) (Dvorak et al., 1992). Vascular hyperpermeability becomes evident within a minute or two of VPF/VEGF injection into normal skin or other tissues and persist for 20 minutes. As Dvorak remembered: "In the Miles assay, Evan's blue dye is injected intravenously into experimental animals, and test substances such as histamine are injected intradermally into the depilated flank skin. Testing cell-free culture supernatants from a variety of human and animal tumor cells in the Miles assay, I found, to my delight, that supernatants from early all generated intense blue spots, whereas those from several normal cells did not. I called this tumor supernatant permeabilizing activity vascular permeability factor of VPF" (Dvorak, 2006). Population-based studies have shown that individuals who are prone to chronic inflammation disorders have an increased risk of cancer development (Table 9.1) (Balkwill et al., 2005).

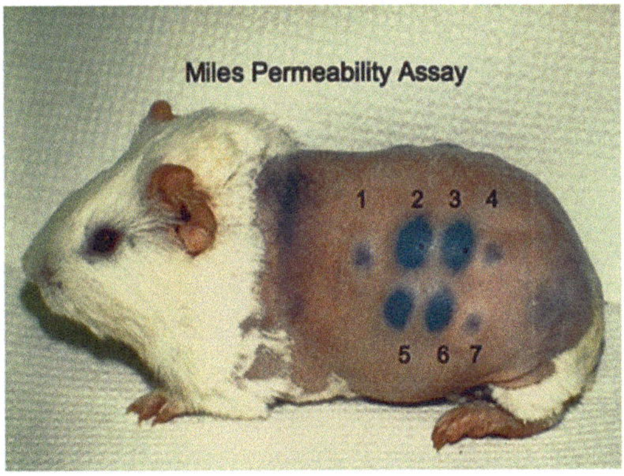

Figure 9.4 Miles permeability assay. Various test substances were injected intradermally into the shaved and depilated flank skin of a Hartley guinea pig, followed immediately by an intravenous injection of Evan's blue dye. The animal was photographed 30 min later. Injected materials were as follows: 1, Neutralizing antibody against VEGF-A; 2 and 5, ascites tumor-associated VEGF-A; 3 and 6, ascites tumor associated VEGF-A plus control immunoglobulin; 4 and 7, ascites tumor-associated VEGF-A plus specific VEGF-A neutralizing antibody. *(Reproduced from Nagy, Benjamin, Zeng, Dvorak, & Dvorak, 2008).*

A meta-analysis found that in ~15 per cent of cancers, tumor initiation can be attributed to infection by viruses, bacteria, and parasites (de Martel et al., 2012). Chromosomal genetic instability in susceptible cell population can be induced by different factors such as reactive oxygen species released from inflammatory cells causing accelerated accumulation of genetic and epigenetic alterations that affect the expression or function of proto-oncogenes and tumor suppressor genes in somatic cells (Coussens & Werb, 2002). Chronic inflammation induces epigenetic changes in the inflamed tissue through changes in the metabolism and the induction of DNA damage (Kominsky, Campbell, & Colgan, 2010). Under a variety of inflammatory conditions, both innate and adaptive immune cells are capable of polarization into their "tumoricidal" (growth arresting) or "tumorigenic" (growth promoting) forms.

The transition from a premalignant lesion to cancer is characterized by a shift of an immune context formed by Th1, Th17, T effector cells, and M1 macrophages to a microenvironment where Th2 cells, Tregs, M2 macrophages, and myeloid derived stem cells (MDSCs) are prevalent. In chronic inflammatory conditions, constant release of colony stimulating factors (CSFs) exacerbate myelopoiesis and interfere with the differentiation of the newly generated cells. As a consequence, poorly differentiated myeloid cells accumulate in the secondary lymphoid organs and tumor tissue itself

(Gabrilovich, Ostrand-Rosenberg, & Bronte, 2012). Treatment with nonsteroidal anti-inflammatory agents decreases the incidence and mortality of several tumor types and low-dose aspirin could reduce the relative risk of cancer mortality (Koehne & Dubois, 2004). Moreover, there is a strong evidence linking carcinogenesis to inflammatory response and reactive oxygen and nitrogen species, and therapeutic strategies for cancer prevention using free radicals and proinflammatory signaling inhibitors have been evaluated in animal models (Rao et al., 2002). Inflammatory cells act in concert with tumor cells, stromal cells, and endothelial cells to create a microenvironment that is critical for the survival, development, and diffusion of the neoplastic mass. These synergies represent important mechanisms for tumor development and metastasis by providing an efficient vascular supply and an easy escape pathway. The most aggressive human cancers, such as malignant melanoma, breast carcinoma, and colorectal adenocarcinoma, are associated with a dramatic host response composed of various inflammatory cells. Stromal events may modulate the response of cancer cells to therapeutic interventions targeting the tumor cells themselves, in particular chemotherapy, and may contribute to the emergence of resistance to therapy. Tumor escape and progression toward metastasis during or after therapy is due to the appearance of resistant tumor cells through a combination of therapy-induced genetic instability and subsequent clonal selection of the most fitted cells.

9.2 Neutrophils

There are two types of neutrophils present in the circulation: circulating neutrophils and marginated neutrophils, which are bound to the endothelium of small blood vessels (Tazzyman, Lewis, & Murdoch, 2009). Neutrophils from the circulation are recruited to the site of inflammation and tumor growth (Friedman, 2002). Neutrophils can kill tumor cells through the release of proteases, membrane perforating agents, reactive oxygen species, and cytokines, such as tumor necrosis factor alpha (TNFα) and interleukin 1-beta (IL-1β) (Di Carlo et al., 2001). Similar to tumor associated macrophages (TAMs), tumor associated neutrophils can be divided into N1 and N2 phenotypes that inhibit or promote cancer development, respectively. Transforming growth factor beta (TGFβ) polarizes neutrophils to an N2 protumor phenotype, whereas inhibition of TGFβ results in a shift to the N1 phenotype (Fridlender et al., 2009). N1 neutrophils increase the activation status of cytotoxic CD8$^+$ T cells and dendritic cells (Fridlender et al., 2009), whereas N2 neutrophils promote cancer by producing angiogenic factors and extracellular matrix (ECM)-degrading enzymes (Nozawa, Chiu, & Hanahan, 2006; Shojaei, Singh, Thompson, & Ferrara, 2008) and by suppressing the antitumor immune response (Schmileau & Finn, 2011).

Neutrophils are the source of several cytokines, including TNF-α, interleukin-1 beta and -12 (IL-1β and IL-12), VEGF, and chemokines, including C-X-C motif chemokine-1, -8, -10 (CXCL-1, -8, -10), C-C motif ligand-3-4 (CCL-3, 4), involved in angiogenesis (Scapini et al., 2000). In breast cancer, release by tumor neutrophils of oncostatin M induces VEGF production from cancer cells (Queen, Ryan, Holzer, Keller-Peck, & Jorcyk, 2005). Expression of human papilloma virus (HPV) 16 in transgenic mice induces a multistage transformation of basal keratinocytes to squamous carcinoma, infiltration of neutrophils and mast cells, activated to secrete matrix metalloproteinase-9 (MMP-9), and angiogenic switch (Coussens & Werb, 1996). In another experimental model of pancreatic carcinogenesis, Rip-Tag2, neutrophils express MMP-9 and their depletion reduced the angiogenic switch of premalignant lesions (Nozawa et al., 2006). Moreover, in knock-out mice for MMP-2 and MMP-9, the absence of both MMP-9-positive neutrophils and MMP-2-expressing-stromal cells resulted in the absence of tumor vascularization and invasion (Masson et al., 2005). Deryugina et al. (2014) demonstrated that tumor associated neutrophils are the major source of MMP-9 in the tumor microenvironment. Neutrophils may facilitate tumor cell extravasation by secretion of MMPs that increase vascular permeability (Coffelt, Wellenstein, & De Visser, 2016). Expression of the hematopoietic cytokines granulocyte-colony stimulating factor (G-CSF) and/or granulocyte macrophage-colony stimulating factor (GM-CSF) was associated with neutrophil recruitment and angiogenesis in malignant HaCaT tumor cells (Gutschalk, Herold-Mende, Fusenig, & Mueller, 2006; Obermueller, Vosseler, Fusenig, & Mueller, 2004). Neutrophil elastase can inhibit angiogenesis by degrading VEGF and fibroblast growth factor-2 (FGF-2) and cleave plasminogen in angiostatin (Ai et al., 2007). Increased recruitment of neutrophils during anti-VEGF therapy promotes tumor progression and treatment resistance (Shojaei et al., 2007). Some metastatic cancer cells can stimulate neutrophils to form neutrophil extracellular traps (NETs), formed from released DNA and ECM molecules intended to trap pathogens in response to infection, which act to support metastatic colonization of secondary organs (Park et al., 2016).

9.3 Eosinophils

Mature eosinophils are characterized by their granules, which have a crystalloid core that binds to the dye eosin. Generally, they are only weakly phagocytic, ingesting some bacteria following activation, but are less efficient than neutrophils at intracellular killing. Eosinophils play a crucial role in allergic reactions and asthma. They are also involved in responses against parasites, in autoimmune and neoplastic diseases, and in fibroses. Tumor-associated eosinophilia has been observed (Ai et al., 2007; Ishibashi et al., 2006; Nielsen et al., 1999). Intravenous transfer of eosinophils can induce tumor vessel normalization and polarize TAMs to M1 phenotype and eosinophils secrete various chemoattractants that enhance the migration of cytotoxic CD8+ T cells into the tumor and are involved in cancer rejection (Carretero et al., 2015). Eosinophils produce a variety of proangiogenic factors, including VEGF, CCL-11, FGF-2, CXCL-8, GM-CSF, platelet derived growth factor (PDGF), TGF-β, and MMP-9 (Munitz & Levi-Schaffer, 2004; Ohno et al., 1997; Puxeddu et al., 2005; Salcedo et al., 2001). Using in vivo models, such as aorta ring and chick embryo chorioallantoic membrane (CAM) assay, eosinophils induce angiogenesis (Fig. 9.5) (Puxeddu et al., 2005; Salcedo et al., 2001).

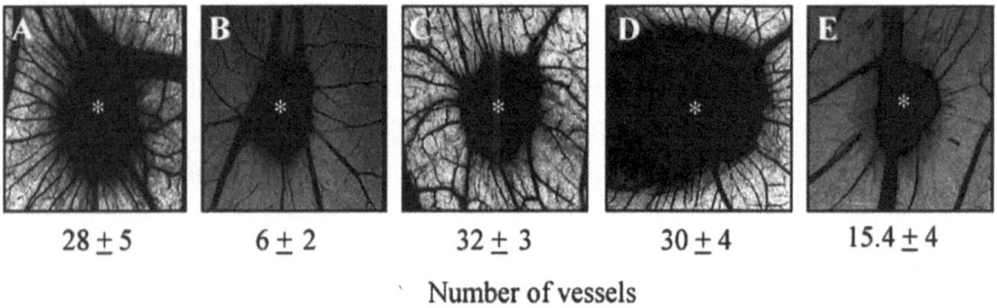

Figure 9.5 Eosinophils induce an angiogenic response in the chick embryo chorioallantoic membrane (CAM). Eosinophil sonicate (A), medium alone (B), cell suspensions of rat mast cells (C), VEGF (D), or eosinophil sonicate preincubated with anti-VEGF neutralizing antibodies (E) were delivered on the top of the CAM, by using a gelatin sponge implant (asterisk). In (A, C, and D), gelatin sponges are surrounded by allantoic vessels that develop radially toward the implant in a "spoked-wheel" pattern. In (B), no vascular reaction is detectable and in (E) reduced vascular reaction is detectable around the sponge. *(Reproduced from Puxeddu, I., Alian, A., Piliponsky, A. M., Ribatti, D., Panet, A., & Levi-Schaffer, F. (2005). Human peripheral blood eosinophils induce angiogenesis.* The International Journal of Biochemistry and Cell Biology, 37*(3), 628–636. https://doi.org/10.1016/j.biocel.2004.09.001).*

9.4 Dendritic cells

These cells derive their name from their morphology characterized by long arborizing cytoplasmic processes of dendrites extending from the cell body for a long distance. Conventional dendritic cells exposed to tumor-derived factors polarize into regulatory dendritic cells that suppressed the proliferation of preactivated T cells (Shurin, Ma, & Shurin, 2013). Dendritic cells may contribute to angiogenesis by secretion of proangiogenic factors or transdifferentiation into endothelial-like cells. Riboldi et al. (2005) demonstrated that alternatively activated dendritic cells produce VEGF. Incubation of dendritic cells with VEGF and oncostatin M led to their transdifferentiation into endothelial-like cells, expressing factor VIII-related antigen, VE-cadherin, and forming vascular-like tubes on Matrigel (Gottfried et al., 2007). Dendritic cells differentiated in the presence of adenosine express higher levels of VEGF, IL-6, IL-8, TGF-β, thus sustaining tumor angiogenesis (Novitskiy et al., 2008). VEGF, IL-6, and TGF-β contribute to reduction of mature dendritic cells and accumulation of immature tolerant dendritic cells and polarization of dendritic cells toward Th2 or Treg induction, expression of activation of immunosuppressive mechanisms (Shurin et al., 2013).

Dendritic cell migration to lymph nodes is a critical event for the efficiency of dendritic cell vaccines in a mouse model and in glioblastoma patients (Mitchell et al., 2015). TNF-α and CXCL-8 produced by immature dendritic cells in ovarian cancer ascites promote angiogenesis in vivo (Curiel et al., 2004) and osteopontin (OPN) released from immature dendritic cells acts as a trigger for the production of the proangiogenic factor IL-1β from monocytes (Naldini et al., 2006). A population of $CD45^+$ $CD11^+$ MHC-II^+ dendritic cells infiltrating human ovarian carcinoma and expressing VE-cadherin assemble into vascular tubes following implantation of tumor-derived CD-45^+-VE-$cadherin^+$ cells in immunodeficient mice (Conejo-Garcia et al., 2004). Treatment with bevacizumab or sorafenib can restore the differentiation of monocytes into dendritic cells (Alfaro et al., 2009).

9.5 Lymphocytes

Lymphocytes consist of a heterogeneous population of cells that differ greatly from each other in terms of origin, life span, areas of settlement within the lymphoid organs, surface structure, and function. Lymphocytes are responsible for the specific immune recognition of pathogens, so they initiate adaptive immune responses. Patients with chronic suppression of adaptive immunity (e.g., patients with acquired immune deficiency syndrome (AIDS) or organ transplants) have an increased risk of developing virus-associated cancers, such as human herpes virus-8 associated Kaposi sarcoma, Epstein Barr Virus (EBV)-associated non-Hodgkin lymphoma, and human papilloma virus (HPV)-associated squamous cell carcinoma (De Visser, Korets, & Coussens, 2005). Otherwise, there is also evidence that adaptive immunity plays a role in promoting cancer, and immune-suppressed patients have diminished risk for solid tumors of epithelial origin, including breast and prostate cancer (De Visser et al., 2005). Tertiary lymphoid structures have been described in close vicinity of tumor mass, as well as within the tumor mass (Dieu-Nosjean, Goc, Giraldo, Sautès-Fridman, & Fridman, 2014) exhibiting a tumor-promoting or tumor-inhibiting role (Finkin et al., 2015; Hiraoka, Ino, & Yamazaki-Itoh, 2016).

Infiltration of tumors by subsets of T lymphocytes can be beneficial and retard tumor growth. T cell infiltration has been correlated with improved prognosis in colorectal cancer, breast cancer, renal cell carcinoma, melanoma, ovarian, and gastrointestinal stroma tumors (GISTs) (Gajewski, Schreiber, & Fu, 2013). Congenital or acquired immunodeficiencies are associated with increased prevalence of malignant diseases. T lymphocytes comprise more than 50 per cent of intratumoral cells and have two major subgroups: CD4, which are considered helper lymphocytes (Th) with $CD4^+$, $CD25^+$, $FoXP3^+$ phenotype, able to activate natural killer (NK) cells and antigen-presenting cells, and $CD8^+$ cells, considered as cytotoxic cells with antitumor effect, which affect apoptosis (Allen & Louise Jones, 2010). A significant reduction in $CD3^+$ T cells between bulky (patients with bulky disease are defined by the presence of a large nodal tumor mass > 10 cm or mediastinal disease) and nonbulky groups has been demonstrated in human diffuse large B cell lymphoma (Fig. 9.6) (Marinaccio et al., 2016). During interaction with dendritic cells, activated $CD4^+$ T cells can acquire neuropilin-1, a coreceptor that binds VEGF from dendritic cells. The resulting neuropilin-1^+ T cells bind dendritic cell-secreted VEGF-A and could behave as VEGF-carrying cells, promoting angiogenesis (Bourbié-Vaudaine, Blanchard, Hivroz, & Roméo, 2006; Ostroumov, Fekete-Drimusz, Saborowski, Kühnel, & Woller, 2018).

Th cells are divided into Th1 cells, which express interferon gamma (IFNγ) and TNFα, and may prevent tumors from growing locally

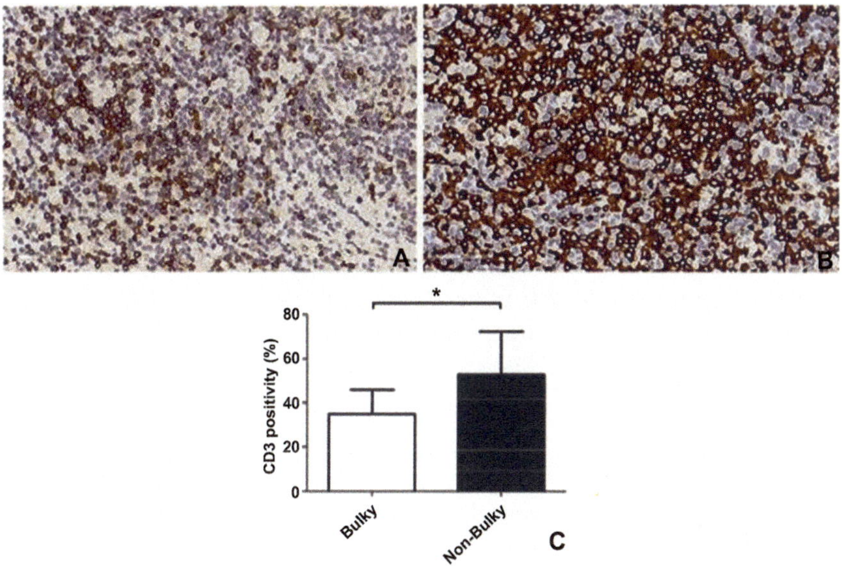

Figure 9.6 CD3 expression in bulky and nonbulky disease groups. (A) Representative image of CD3 expression in a case with bulky disease. (B) Representative image of CD3 expression in a case with nonbulky disease. (C) Comparison between bulky and nonbulky disease groups with a significant difference between the groups in the expression of CD3. *$P < .05$. *(Reproduced from Marinaccio, C., Ingravallo, G., Gaudio, F., Perrone, T., Ruggieri, S., Opinto, G., et al. (2016). T cells, mast cells and microvascular density in diffuse large B cell lymphoma.* Clinical and Experimental Medicine, 16*(3), 301–306. https://doi.org/10.1007/s10238-015-0354-5).*

(Fridman, Pagès, Sauts-Fridman, & Galon, 2012), and Th2 cells, generally, protumor cells, which express IL-4, IL-5, and IL-13 (Ivanova & Orekhov, 2015). T regulatory cells (Tregs) are a subset of $CD4^+/CD25^+$, $FoXP3^+$ cells, which are generated by antigenic stimulation, such as infection and cancer (Allen & Louise Jones, 2010). Tregs are involved in the suppression of tumor immunity through their recruitment in the tumor microenvironment mediated by VEGF, hypoxia, and CCL-33/CCL-17/CC-4 axis (Gobert et al., 2009; Liu, Workman, & Vignali, 2016). In ovarian cancer, Tregs secrete higher amounts of VEGF-A and promote endothelial cell proliferation in vitro and in vivo (Facciabene et al., 2011). Depletion of Tregs using an anti-CD25 antibody in combination with adaptive transfer of tumor-specific cytotoxic T lymphocytes (CTLs) was more effective than each single therapy in a syngeneic murine renal carcinoma model (Ohmura et al., 2008). VEGF blocks cytotoxic T cell trafficking and activity by modulating the inhibitory checkpoints of T cells, and inhibiting VEGF expression from T cells induces vessels normalization (Palazon et al., 2017). Antitumor $CD4^+$ T cells modulate angiogenic gene expression in tumors, leading to vessel normalization (De Palma & Jain, 2017).

9.6 Natural killer cells

NK cells originate from hematopoietic stem cells (HSCs) and undergo maturation primarily in the bone marrow, where stromal cells produce factors sustaining proliferation and differentiation of NK progenitors (Miller, Verfaillie, & McGlave, 1992). Immature NK cells are $CD16^-$ and $CD56^-$, but express CD161 and exert cytolytic functions (Bennett et al., 1996). During differentiation process, the expression of CD34 is progressively lost while other surface antigens specific for the NK lineage appear. In addition, NK cells progressively acquire the expression of inhibitory and activating receptors as well as their functions of cytokine secretion and cytotoxicity (Moretta et al., 2001). NK cells account for 5–15 per cent of the peripheral blood lymphocytes and exert cytotoxic functions (Trinchieri, 1989). More than 90 per cent of NK cells are found in the blood circulation, where they express high amounts of CD16 and low amounts of CD56, while NK cells express high amounts of CD56, lack of CD16, and have low cytotoxic activity (Freud & Caligiuri, 2006). Less than 10 per cent of NK cells are found in healthy tissues, such as skin, gut, spleen, liver, lungs, and uterus during pregnancy. Under pathological conditions and during inflammation, NK cells extravasate into the lymph nodes and accumulate at sites of tumor growth.

Mice with compromised NK cell function are more susceptible to carcinogen-induced cancers, and individuals lacking NK cells suffer from persistent viral infections and as a consequence die prematurely. NK cells exert cytotoxic functions mainly directed against virus-infected cells and tumor cells (Trinchieri, 1989). They also display immunomodulatory functions through secretion of immunoregulatory cytokines, like IFN-γ, TNF-α, IL-10, and IL-13, and GM-CSF, but they have a low natural cytotoxicity (Cooper et al., 2001). NK cells spontaneously lyse tumor targets in vivo and in vitro without requiring immunization or preactivation. Cytokines including IFN-α/-β, IL-2, IL-12, and IL-15 enhance NK cell–mediated cytotoxicity, whereas IL-2 and IL-12 induce NK proliferation, and IL-1, IL-2, IL-12, IL-15, IL-18, and TNFα induce NK lymphokine production.

NK cells provide surveillance against tumor cells and virus infected cells (Hayakawa & Smyth, 2006). In experimental in vivo models, NK cells can destroy tumor cells, and are effective in eliminating tumor cells and preventing metastasis (Smyth, Hayakawa, Takeda, & Yagita, 2002). Higher incidence lymphoproliferative disorders have been demonstrated in patients with Chédiak–Higashi syndrome, which have profound deficits in NK activity, and in patients with X-linked lymphoproliferative diseases, NK activity is deficient as well (Sullivan & Woda, 1989). NK cells often are decreased in pathological conditions, including cancer and AIDS (Trinchieri, 1989). NK cells secrete proangiogenic cytokines, including VEGF, placental growth factor (PlGF), IL-8, IL-10, TGFβ, angiopoietin-1 and angiopoietin-2 (Ang-1 and Ang-2). Decidual NK cells are able to release angiogenic factors, including VEGF, PlGF, and IL-8, necessary for spiral artery formation during decidualization (Vacca, Moretta, Moretta, & Mingari, 2011). Loss of hypoxia inducible factor 1 alpha (HIF-1α) in NK cells increased the bioavailability of VEGF by decreasing the infiltration of NK cells that express angiostatic soluble VEGF receptor-1 (VEGFR-1) (Krzywinska et al., 2017). Moreover, NK cell depletion leads to reduction of corneal angiogenesis associated with reduced macrophage infiltration and lower levels of VEGF-A (Lee et al., 2014), and zoledronic acid synergizes with IL-2 in inducing proangiogenic activity of NK cells acting on VEGF (Hoeres et al., 2018).

Several studies have shown that tumor infiltrating NK (TINK) and tumor-associated NK (TANK) are compromised in their ability to lyse tumor cells; they are potentially protumorigenic and can also acquire a proangiogenic phenotype (Bruno, Ferlazzo, & Albini, 2014). A protumorigenic phenotype of cancer infiltrating NK cells has been demonstrated due to their ability to release angiogenic factors and immunosuppressive cytokines (Bruno et al., 2014). In non–small cell lung cancer (NSCLC) NK cells synthesize higher VEGF, IL-8, and PlGF as compared to controls (Bruno et al., 2013). Angiogenin, MMP-2, and tissue inhibitor of MMPs (TIMPs) are expressed by NK cells in patients with colorectal cancer (Bruno et al., 2018). Moreover, STAT3/STAT5 activation was observed in tumor NK cells, and treatment with pimozide, a STAT5 inhibitor, reduced angiogenic capability of endothelial cells, inhibiting VEGF and angiogenic production (Bruno et al., 2018). NK cells are required mediators of angiogenesis inhibition by IL-12 in an IFN-γ-dependent manner, and NK cell cytotoxicity of endothelial cells is a potential mechanism by which IL-12 can suppress neovascularization (Yao et al., 1999). IL-12 receptors indeed are present primarily on NK cells and T cells (Trinchieri, 1993). IL-12 suppresses angiogenesis through NK-mediated cytotoxicity of endothelial cells (Yao et al., 1999) and IL-12-activated lymphocytes exert an antiangiogenic activity through the release of IFN-γ and downmodulation of VEGF (Cavallo et al., 2001). The genetic inactivation of STAT5, which is required for NK-cell-mediated immunosurveillance, upregulates VEGF-A in NK cells and enhances angiogenesis in mouse lymphoma models and on healthy donor-derived NK cells (Gotthardt et al., 2016).

9.7 Platelets

Platelets contain a whole range of pro- and anti-angiogenic compounds, which are endocytosed and sequestered in different populations of alpha granules and accumulation of platelets in some tumors, and release of angiogenic molecules could further stimulate tumor growth (Klement et al., 2009). Tumor cells express tissue factor on their plasma membranes which, in turn, provide a surface for prothrombinase activation; thus, intravascular clotting tumor cells provide the functions played by platelets in intravascular clotting (Dvorak et al., 1983). Moreover, tumor cells shed portions of their plasma membranes in the form of vesicles (exosomes), which possess procoagulant activity (Dvorak et al., 1983). Circulating tumor cells are associated with activated platelets, which secrete protective signals, such as TGFβ, which in turn upregulates NF-kB signaling in circulating tumor cells (Labelle, Begum, & Hynes, 2014).

Platelet–tumor cell interaction was shown to enhance tumor metastasis by stimulating epithelial mesenchymal transition (EMT). Platelets can guide the formation of an early metastatic niche by paracrine signaling between tumor cells platelets and granulocytes. CXCL5/7 chemokine production by tumor cell–activated platelets recruits CXCR2^{+} granulocytes to metastatic tissues to form a microenvironment favorable for metastatic seeding (Labelle et al., 2014). Platelets can facilitate tumor metastasis by releasing bioactive factors (Gay & Felding-Habermann, 2011) and by protecting circulating tumor cells from elimination by NK cells (Palumbo et al., 2005). Therapies interfering with platelet function, as long-term treatment with aspirin, have shown success in reducing risk of metastatic disease, which further supports a metastasis-promoting role of platelets (Rothwell et al., 2012).

9.8 Mast cells

Although some evidence suggest that mast cells can promote tumorigenesis and tumor progression, there are some clinical data as well as experimental tumor models in which mast cells seem to have functions that favor the host (Ribatti & Crivellato, 2009). Mast cells attracted in the tumor microenvironment by stem cell factor (SCF) secreted by tumor cells produce several angiogenic factors as well as MMPs, which promote tumor vascularization and invasiveness, respectively (Ribatti & Crivellato, 2009). Similar to neutrophils and macrophages, two subpopulations of mast cells may exist, one protumorigenic and the other one antitumorigenic. Mast cells are a major source of histamine, which modulate tumor growth through H1 and H2 receptors (Fitzsimons et al., 1997). H1 receptor antagonists significantly improved overall survival rates and suppressed tumor growth as well as the infiltration of mast cells and VEGF levels through the inhibition of HIF-1α expression in B16F10 melanoma-bearing mice (Jeong et al., 2012). Mast cells exert immunosuppression releasing TNF-α and IL-10, which are essential in promoting the immune tolerance mediated by regulatory T (Treg) cells, and stimulate immune tolerance and tumor promotion (Grimbaldeston, Nakae, Kalesnikoff, Tsai, & Galli, 2007; Ullrich, Nghiem, & Khaskina, 2007).

Mast cells may promote inflammation, inhibition of tumor cell growth, and tumor cell apoptosis by releasing cytokines, such as IL-1, IL-4, IL-6, IL-8, monocyte chemotactic protein-3 and protein-4 (MCP-3 and MCP-4), TGF-β, and chymase. Chondroitin sulfate may inhibit tumor cell diffusion, and tryptase causes both tumor cell disruption and inflammation through activation of protease-activated receptors (PAR-1 and PAR-2) (Glowacki & Mulliken, 1982). Within the developing tumor environment, mast cells do not act alone. They are recruited early in tumor development and play a critical role in both angiogenesis and tissue remodeling. As tumor growth progresses, mast cells recruit eosinophils and neutrophils and activate T and B cell immune responses (Kinet, 2007). Data on mast cells and metastasis have been scant and poor. Yano et al. (1999) reported that mast cell infiltration around gastric cancer cells correlated with tumor angiogenesis and metastasis. Increased mast cell number has been correlated with a poor prognosis in human melanoma (Fig. 9.7) (Ribatti, Vacca, & Ria, 2003).

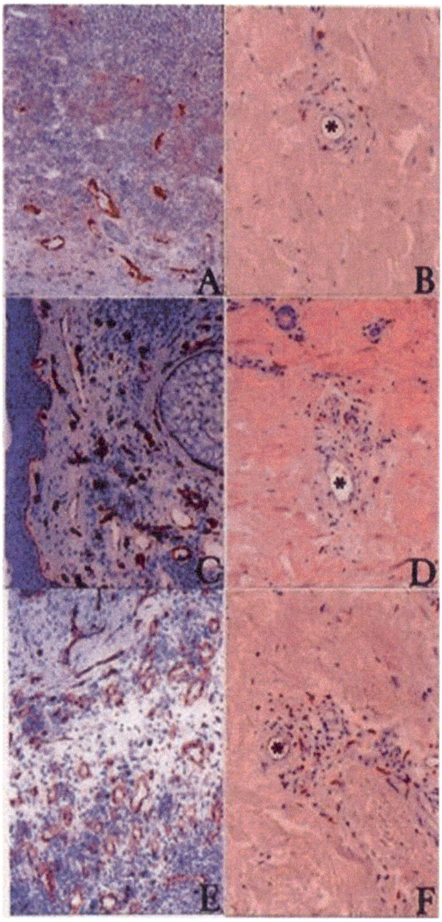

Figure 9.7 Adjacent sections of common nevi (A, B), nevi with varying degrees of melanocytic atypia (ADMA) (C, D), and metastatic (E, F) melanoma stained with CD31 (A, C, E) for microvessels and with tryptase for mast cells (B, D, F). Note the progressive increase of microvessels and mast cells located around blood capillaries (asterisks) from common nevi to nevi with ADMA and from latter to metastatic melanoma. *(Reproduced from Ribatti, D., Vacca, A, & Ria, R. (2003). Neovascularization, expression of basic fibroblast growth factor-2, and mast cells with tryptase activity increase simultaneously with pathological progression in human malignant melanoma.* European Journal of Cancer, *39, 666–675).*

Mast cells produce several proangiogenic factors, including FGF-2, VEGF, IL-8, TNF-α, TGF-β, and nerve growth factor (NGF) (Abdel-Majid & Marshall, 2004; Boesiger et al., 1998; Grützkau et al., 1998; Kanbe et al., 2000; Moller et al., 1993; Nilsson et al., 1997; Qu, Kayton, & Ahmadi, 1998;Qu et al., 1998; Walsh, Trinchieri, Waldorf, Whitaker, &

Murphy, 1991), and migrate in vivo and in vitro in response to VEGF and placental PlGF-1 (Detmar et al., 1998; Detoraki et al., 2009; Gruber, Marchese, & Kew, 1995). Human lung mast cells express VEGF-A, VEGF-B, VEGF-C, and VEGF-D, and supernatants of prostaglandin E2 (PGE2)- and 5′-N-ethylcarboxamido- adenosine (NECA)-activated lung mast cells induced angiogenic response in the chick CAM assay that was inhibited by an anti-VEGF-A antibody (Detoraki et al., 2009). Granulated murine mast cells and their granules are able to stimulate an intense angiogenic reaction in the CAM assay, partly inhibited by anti-FGF-2 and anti-VEGF antibodies (Ribatti et al., 2001). Intraperitoneal injection of the degradating compound 48/80 causes a vigorous angiogenic response in the rat mesentery window angiogenic assay and in mice (Norrby, Jakobsson, & Sörbo, 1986; Norrby, Jakobsson, & Sörbo, 1989). Histamine and heparin stimulate proliferation of endothelial cells in vitro and are angiogenic in the CAM assay (Fig. 9.8) (Ribatti, Roncali, Nico, & Bertossi, 1987; Sörbo, Jakobbson, & Norrby, 1994). Mast cells store in their secretory granules preformed active serine proteases, including tryptase and chymase (Metcalfe, Baram, & Mekori, 1997). Tryptase stimulates the proliferation of endothelial cells, promotes vascular tube formation in vitro, degrades connective tissue matrix, and activates matrix MMPs and plasminogen activator (PA), which in turn degrade the extracellular matrix with consequent release of VEGF or FGF-2 from their matrix-bound state (Blair et al., 1997). The expression of mast cell chymase and tryptase correlated with mast cell maturation and angiogenesis during tumor progression in chemically induced tumor growth in BALB/c mouse (de Souza et al., 2012). Mast cells contain MMPs and tissue inhibitors of MMPs (TIMPs) (Koskivirta et al., 2006; Tanaka, Yamane, & Matsuda, 2001), which intervene in regulation of ECM degradation, allowing release of angiogenic factors. Mast cell—deficient W/Wv mice exhibit a decreased rate of tumor angiogenesis (Starkey, Crowle, & Taubenberger, 1988). Development of squamous cell carcinoma in a human papilloma virus (HPV) 16 infected transgenic mouse model of epithelia carcinogenesis provided experimental support for the early participation of mast cells in tumor growth and angiogenesis (Coussens et al., 1999; Coussens, Tinkle, Hanahan, & Werb, 2000).

Mast cells infiltrated hyperplasia, dysplasias, and the invasive front of carcinomas, but not the core of tumors. Accumulation occurred proximal to developing capillaries and the stroma surrounding the advancing tumor mass (Coussens et al., 1999). Infiltration of mast cells and activation of MMP-9 coincided with the angiogenic switch in premalignant lesions

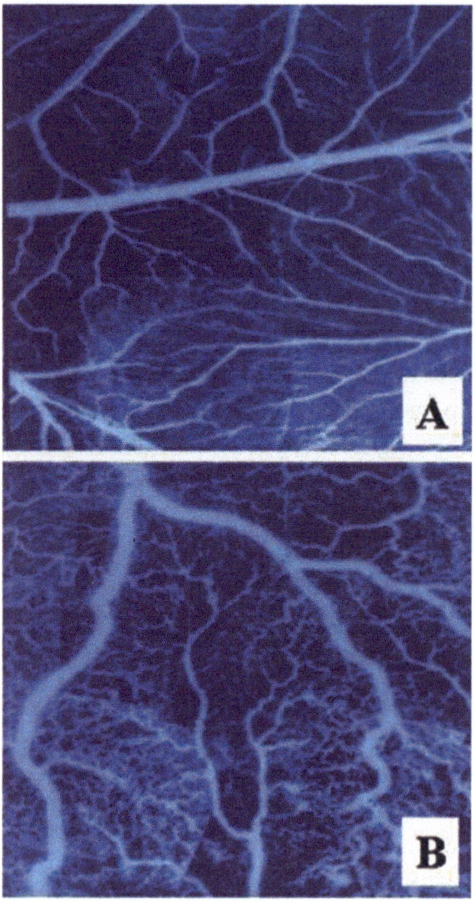

Figure 9.8 Photographic reconstruction of a portion of chick embryo chorioallantoic membrane treated with heparin (B) compared with a control (A). Note that in the control the vessels run straight and interdigitate regularly, whereas in a heparin-treated specimen, the vessel course is irregular or sinuous, their branching is frequent, and vascular loops are numerous. *(Reproduced from Ribatti D, Roncali L, Nico B & Bertossi M (1987). Effects of exogenous heparin on the vasculogenesis of the chorioallantoic membrane. Cell Tissue Organs, 130 (3):257–263. https://doi.org./10.1159/000146454).*

through the release of proangiogenic molecules from the ECM. Remarkably, premalignant angiogenesis was abrogated in a mast cell–deficient HPV 16 transgenic mouse indicating that neoplastic progression in this model involved infiltration of mast cells in the skin (Coussens et al., 1999, 2000). By using the same in vivo transgenic mouse model, it has been demonstrated that genetic elimination of mature T and B lymphocytes limits neoplastic progression (Andreu et al., 2010; De Visser et al., 2005). Moreover, in prostate tumors derived from both tumor transgenic adenocarcinoma of the mouse prostate (TRAMP) mice and human patients, mast cells promote well-differentiated adenocarcinoma growth (Pittoni et al., 2011). An increased number of mast cells have been demonstrated in angiogenesis associated with vascular tumors, like hemangioma and hemangioblastoma), as well as a number of hematological and solid tumors, including lymphomas (Fukushima, Satoh, Sano, & Tokunaga, 2001; Ribatti et al., 1998), multiple myeloma (Fig. 9.9) (Nico, Mangieri, Crivellato, Vacca, & Ribatti, 2008; Ribatti et al., 1999), myelodysplastic syndrome (Ribatti et al., 2002), B-cell chronic lymphocytic leukemia (Molica, Vacca, Crivellato, Cuneo, & Ribatti, 2003), breast cancer (Bowrey et al., 2000; Hartveit, 1981), colorectal cancer (Lachter, Stein, Lichtig, Eidelman, & Munichor, 1995), uterine cervix cancer (Benítez-Bribiesca, Wong, Utrera, & Castellanos, 2001; Graham & Graham, 1966), melanoma (Reed, Scott McNutt, Bogdany, & Albino, 1996; Ribatti et al., 2003), and pulmonary adenocarcinoma (Ullah, Nagi, & Lail, 2012). Indeed, a prognostic significance has been attributed to mast cells and microvascular density also in squamous cell cancer of the esophagus (Elpek et al., 2001). An association of VEGF and mast cells with angiogenesis has been demonstrated in laryngeal carcinoma and in small lung carcinoma (Imada, Shijubo, Kojima, & Abe, 2000; Sawatsubashi et al., 2000; Takanami, Takeuchi, & Naruke, 2000; Tomita, Matsuzaki, & Onitsuka, 2000).

9.9 Monocytes and macrophages

Cells belonging to the monocyte–macrophage lineage are a major component of the leukocyte infiltration in tumors. Macrophages that accumulate at sites of inflammation largely derive from the extravasation of bone marrow–derived monocytes (Perdiguero & Geissmann, 2016). Neuropilin-1 expressing monocytes are involved in tumor vessel normalization, by secreting different cytokines that promote pericyte recruitment leading to enhanced pericyte coverage of tumor vessel (Carrer et al., 2012). Tumor-derived chemoattractants ensure macrophage recruitment, including colony stimulating factor-1 (CSF-1), the CC chemokines CCL-2, CCL-3, CCL-4, CCL-5, and CCL-8, and VEGF secreted by both tumor and stromal elements (Mantovani, Sozzani, Locati, Allavena, & Sica, 2002). Expression of CSF-1 receptor (CSF-1R), also known as macrophage CSF (M-CSF) on TAMs and high CSF-1 levels, correlate with poor prognosis in different solid tumors (Zhang et al., 2012). Monoclonal antibodies and small molecules targeted against CSF-1R and CSF-1 are under evaluation in clinical

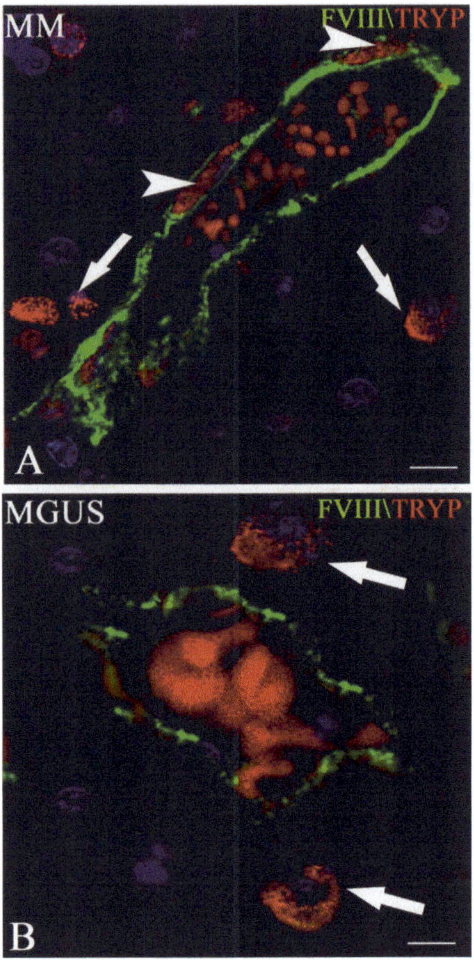

Figure 9.9 Double factor VIII-related antigen (FVIII-RA) (green) and tryptase (red) confocal laser microscopy from multiple myeloma (MM) (A) and monoclonal gammopathy of undetermined significance (MGUS) (B) bone marrow bioptic specimens. In (A) an MM vessel is lined by both endothelial cells positive to FVIII-RA and by mast cells positive to tryptase. Mast cells containing tryptase-positive granules (*arrow*) are also recognizable on the abluminal side of the vessel. In (B) an MGUS vessel is lined only by endothelial cells positive to FVIII-RA and is surrounded by tryptase-positive mast cells (*arrow*). *(Reproduced from Nico, B., Mangieri, D., Crivellato, E., Vacca, A., & Ribatti, D. (2008). Mast cells contribute to vasculogenic mimicry in multiple myeloma.* Stem Cells and Development, 17*(1), 19–22. https://doi.org/10.1089/scd.2007.0132).*

trials as single agent or in combination therapy. Blocking recruitment of TAMs in mammary tumors by treating mice with agents that blocked CSF-1R signaling diminished tumor vascularity, slowed primary tumor development, reduced formation of pulmonary metastases, and improved survival, when given in combination with chemotherapy (DeNardo et al., 2011).

Activated macrophages are generally categorized in two types, M1 (classically activated) and M2 (alternatively activated) (Balkwill et al., 2005). M1 macrophages are able to kill microorganisms as well as tumor cells and secrete high levels of proinflammatory cytokines and tumoricidal agents (TNF-α and IL-12), IL-1β, IL-23, the IFNγ inducible cytokines CXCL-9, CXCL-10, CXCL-11, reactive nitrogen and oxygen intermediates (RNI and ROI) (Balkwill et al., 2005). These cytokines are produced in response to microbial stimuli, i.e., lipopolysaccharide. M1 macrophages are involved in polarized TH1 response and mediate resistance against intracellular parasites and tumors. TAMs are generally more abundant in the stromal areas of tumors, but also infiltrate both viable and necrotic cancer cell areas. In the tumor microenvironment, TAMs are mainly constituted by M2 elements (Mantovani et al., 2002), which have poor attitude to destroy tumor cells but are better adapted to promoting angiogenesis, repairing and remodeling wounded or damaged tissues, and suppressing adaptive immunity (Sica, Schioppa, Mantovani, & Allavena, 2006). In hypoxic tumor regions, M2 macrophages predominate (Engblom, Pfirschke, & Pittet, 2016), while in well-perfused tumor regions M2 macrophage genes are low expressed (Abramovitch et al., 1998; Huang et al., 2012). The interaction between CSF-1R and CSF-1 is crucial for differentiation and proliferation of M2 macrophages (Chitu & Stanley, 2006). Macrophages are involved in TH2 reactions, promote killing and encapsulation of parasites, tumor progression, tissue repair, and remodeling and have immunoregulatory functions. In regressing and nonprogressing tumors, TAMs mainly resemble the M1 type and exhibit antitumor activity, while in malignant and advanced tumors, TAMs are biased toward the M2 phenotype that favors tumor malignancy (Qian & Pollard, 2010). Unique cell surface markers that distinguish the two TAM phenotypes remain elusive and expression of M1/M-2 associated molecules is highly dependent on tumor type, tumor stage, intratumoral localization, hypoxia, and other microenvironmental signals (Mantovani et al., 2002).

The phenotype of polarized M1-M2 macrophages has the potential to be reversed (Guiducci, Vicari, Sangaletti, Trinchieri, & Colombo, 2005). Rolny et al. (2011) demonstrated that the host-produced histidine-rich glycoprotein (HRG) skewed TAM polarization away from the M2 to a tumor-inhibiting M1-like phenotype. In this context, HRG promoted antitumor immune responses and vessel normalization. Extensive TAM infiltration correlates with a poor prognosis for breast, prostate, cervix, and bladder cancer patients (Talmadge & Gabrilovich, 2013). TAMs produce several proangiogenic cytokines, including VEGF, TNF-α, IL-8, and FGF-2, as well as extracellular matrix-degrading enzymes, including MMP-2, MMP-7, MMP-9, -12, and cycloxygenase-2 (COX-2) (Klimp et al., 2001; Naldini & Carraro, 2005). In an experimental model of subcutaneous melanoma, both angiogenesis and growth rate correlate with tumor infiltration by macrophages expressing angiotensin-1 receptor and VEGF (Egami et al., 2003). Lewis lung carcinoma cells expressing IL-1-β develop neovasculature with macrophage infiltration and enhance tumor growth in wild-type but not in MCP-1-deficient mice, suggesting that macrophage involvement might be a prerequisite for neovascularization and tumor progression (Nakao et al., 2005). In a murine model of mammary carcinoma, deficiency of macrophage colony-stimulating factor (M-CSF), an inductor of macrophage recruitment in tumor tissues, reduces progression to invasive carcinoma and metastasis (Lin, Nguyen, Russell, & Pollard, 2001). In polyoma middle-T(PyMT)-induced mouse mammary tumors, accumulation of macrophages in premalignant lesions precedes the angiogenic switch and the progression into invasive tumors (Lin et al., 2006, 2007). Inhibition of macrophage homing into the tumor microenvironment delayed the angiogenic switch, whereas genetic restoration of macrophages rescued the vascular phenotype. Upregulation of angiogenesis in TAMs is stimulated by hypoxia and acidosis (Bingle, Brown, & Lewis, 2002). Activated macrophages synthesize and release inducible nitric oxide synthase, which increases blood flow and promotes angiogenesis (Jenkins et al., 1995). Angiogenic factors secreted by macrophages stimulate migration of other accessory cells that potentiate angiogenesis, in particular mast cells (Gruber et al., 1995). OPN deeply affects the proangiogenic potential of human monocytes (Denhardt, Noda, O'Regan, Pavlin, & Berman, 2001), and may affect angiogenesis by acting directly on endothelial cells and/or indirectly via mononuclear phagocyte engagement, enhancing the expression of TNF-α and IL-1-β in mononuclear cells (Leali et al., 2003; Naldini et al., 2006). Macrophages release IL-12, which causes tumor regression and reduces metastasis in animal models, through the promotion of antitumor immunity and also to the inhibition of angiogenesis (Colombo & Trinchieri, 2002). The antiangiogenic activity is mediated by IFN-γ production, which in turn induces the chemokine IFN-γ-inducible protein-10 (Angiolillo et al., 1995; Romagnani et al., 2001). Moreover, IL-12 inhibits VEGF production by breast cancer cells and regulates stromal cell interactions, leading to decreased MMP-9 and increased tissue inhibitor of TIMP-1 production (Dias, Boyd, & Balkwill, 1998).

Upregulation of angiogenesis in TAMs is stimulated by hypoxia and acidosis (Bingle et al., 2002). Mice deficient in HIF-2α in myeloid cells displayed reduced TAM infiltration in both murine hepatocellular and colitis-associated colon carcinoma models (Imtiyaz et al., 2010). Mouse mammary tumors, exhibiting enriched M2-like TAMs in hypoxic tumor areas, demonstrated an increased angiogenic activity (Movahedi et al., 2010). The developing vasculature in tumors lacking myeloid-cell-derived VEGF-A was less tortuous, with increased pericyte coverage (indicating increased maturation), decreased vessel length, with evidence of vascular normalization and increased susceptibility to chemotherapeutic agents (Stockmann et al., 2008). A significant relationship between the number of TAMs and the density of blood vessels has been established in human tumors, including breast carcinoma, melanoma, glioma, squamous cell carcinoma of the esophagus, bladder carcinoma, and prostate carcinoma (Hanada et al., 2000; Koide, Nishio, Sato, Sugiyama, & Miyagawa, 2004; Leek, Landers, Harris, & Lewis, 1999; Lissbrant et al., 2000; Nishie et al., 1999). We have investigated CD68 expression and its relationship with microvascular density in chemoresistant and chemosensitive patients affected by diffuse large B-cell lymphoma (DLBCL) and observed that CD68 expression as well as microvascular density were increased in chemoresistant patients when compared with chemosensitive patients (Fig. 9.10) (Marinaccio et al., 2014). Depletion of TAMs reduces to about 50 per cent tumor vascular density, leading to areas of necrosis by loss of blood supply within the tumor mass, and macrophages accumulate particularly in such necrotic and hypoxic areas in different neoplasia, like human endometrial, breast, prostate, and ovarian carcinomas (Chen et al., 2009; Leek et al., 1999; Ohno et al., 2004). Chen et al. (2009) demonstrated that monocytes induce vascular endothelial cell gene expression and develop tubelike structures when they are cultured with the bone marrow from patients with MM that express an angiogenic factor, namely

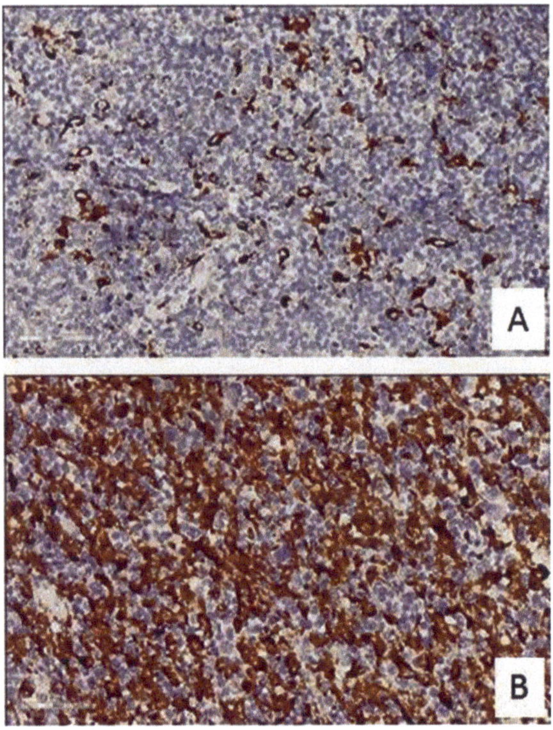

Figure 9.10 Macrophages stained with an anti-CD68 antibody expression in responders (A) and nonresponders (B) in diffuse large B cell lymphoma patient groups. Note a significant difference between the groups in the expression of CD68. *(Reproduced from Marinaccio, C., Ingravallo, G., Gaudio, F., Perrone, T., Nico, B., Maoirano, E., et al. (2014). Microvascular density, CD68 and tryptase expression in human diffuse large B-cell lymphoma.* Leukemia Research, *38(11), 1374–1377. https://doi.org/10.1016/j.leukres.2014.09.007).*

pleiotrophin; the tubulogenesis was specifically blocked by antipleiotrophin antibodies. Moreover, when coinjected with human MM cells into severe combined immunodeficient (SCID) mice, green fluorescent protein-marked human monocytes were found to be incorporated into tumor blood vessels and expressed human vascular endothelial cell protein markers and genes that were blocked by anti-pleiotrophin antibodies.

In patients with active MM, fluorescence activated cell sorting (FACS) analysis on isolated bone marrow mononuclear cells revealed higher percentages of CD68^{+} macrophages than in patients with nonactive disease or those with MGUS (Scavelli et al., 2007). Bone marrow macrophages in patients with active MM were similar to paired endothelial cells and contributed to angiogenesis through a vasculogenic mimicry (Scavelli et al., 2007). When macrophages from these patients were exposed in vitro to VEGF and FGF-2, they transformed into cells functionally and phenotypically similar to endothelial cells because they were able to generate capillary-like networks (Scavelli et al., 2007). Moreover, bone marrow biopsies of active MM patients examined on confocal laser microscopy displayed macrophages with both endothelial cell-like (i.e., CD68/FVIII-RA double positive) and apparently typical (i.e., CD68 positive/FVIII-RA negative) features located in the microvessel wall. Figures of this type were rare in nonactive MM patients and absent in MGUS (Fig. 9.11) (Scavelli et al., 2007). TAMs are associated with high lymphatic vessel density and lymphatic metastasis. TAMs secrete prolymphangiogenic factors (Jones, Pereira, & Padera, 2018) and VEGF-C acts as a chemotactic factor for VEGFR-3 expressing macrophages (Skobe et al., 2001). VEGFR3+ TAMS were recruited to breast and lung tumors in response to paclitaxel and mediated chemotherapy resistance through VEGF-C production and lymphangiogenesis (Alishekevitz et al., 2016).

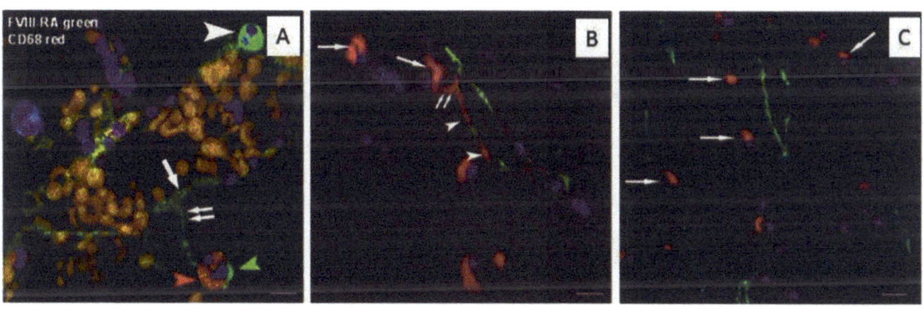

Figure 9.11 Microvessel lines by flattened FVIII-RA positive multiple myeloma endothelial cells (*arrow*), an FVIII-RA positive macrophage (*arrowhead*) showing protrusions connected to multiple myeloma endothelial cells, and another macrophage containing double-labeled CD68 (*red arrowhead*) and FVIII-RA (*green arrowhead*) granules in the cytoplasm and connected to multiple myeloma endothelial cells by an FVIII-RA positive cytoplasmic protrusion (*double arrow*). Erythrocytes (orange) are well recognizable inside the lumen (B). Another microvessel formed by FVIII-RA positive (green) multiple myeloma endothelial cells and CD68 positive (red, *arrowheads*) tracts that belong to the cytoplasmic protrusions (*double arrow*) of macrophages, some of which are arrowed (C). MGUS microvessels are formed by only FVIII-RA positive endothelial cells: macrophages (*arrows*) are randomly scattered in the tissue and independent of them. *(Reproduced from Scavelli C, Nico B, Cirulli T , Ria R, Di Pietro G, Mangieri D, et al. (2007). Vasculogenic mimicry by bone marrow macrophages in patients with multiple myeloma. Oncogene 27 (5):663–674. https://doi.org/10.138/sj.onc.1210691).*

9.10 Note

1. Sir Frank Macfarlane Burnet assumed that in a population of mesenchymal cells there are a certain number of different clones and that it is characteristic of any such clone that is composed of cells which are immunologically competent toward a certain antigenic determinant. If the antigenic determinant makes effective contact with such a cell, it is stimulated to proliferate (clonally expand) and under appropriate circumstances some of the descendant cells will be converted into plasma cells and produce antibodies that combine with the antigen. If the antigen is part of the surface of a virus or bacterium, then the antibody labels that organism as foreign ("not-self").

References

Abdel-Majid, R. M., & Marshall, J. S. (2004). Prostaglandin E2 induces degranulation-independent production of vascular endothelial growth factor by human mast cells. *Journal of Immunology, 172*(Issue 2), 1227–1236. https://doi.org/10.4049/jimmunol.172.2.1227. American Association of Immunologists.

Abramovitch, R., Marikovsky, M., Meir, G., & Neeman, M. (1998). Stimulation of tumour angiogenesis by proximal wounds: Spatial and temporal analysis by MRI. *British Journal of Cancer,* 77(3), 440–447. https://doi.org/10.1038/bjc.1998.70.

Ai, S., Cheng, X. W., Inoue, A., Nakamura, K., Okumura, K., Iguchi, A., et al. (2007). Angiogenic activity of bFGF and VEGF suppressed by proteolytic cleavage by neutrophil elastase. *Biochemical and Biophysical Research Communications, 364*(2), 395–401. https://doi.org/10.1016/j.bbrc.2007.10.027.

Alfaro, C., Suarez, N., Gonzalez, A., Solano, S., Erro, L., Dubrot, J., et al. (2009). Influence of bevacizumab, sunitinib and sorafenib as single agents or in combination on the inhibitory effects of VEGF on human dendritic cell differentiation from monocytes. *British Journal of Cancer, 100*(7), 1111–1119. https://doi.org/10.1038/sj.bjc.6604965.

Alishekevitz, D., Gingis-Velitski, S., Kaidar-Person, O., Gutter-Kapon, L., Scherer, S. D., Raviv, Z., et al. (2016). Macrophage-induced lymphangiogenesis and metastasis following paclitaxel chemotherapy is regulated by VEGFR3. *Cell Reports, 17*(5), 1344–1356. https://doi.org/10.1016/j.celrep.2016.09.083.

Allen, M., & Louise Jones, J. (2010). Jekyll and Hyde: The role of the microenvironment on the progression of cancer. *The Journal of Pathology, 223*(2), 163–177. https://doi.org/10.1002/path.2803.

Andreu, P., Johansson, M., Affara, N. I., Pucci, F., Tan, T., Junankar, S., et al. (2010). FcRgamma activation regulates inflammation-associated squamous carcinogenesis. *Cancer Cell, 17*(2), 121–134. https://doi.org/10.1016/j.ccr.2009.12.019.

Angiolillo, A. L., Sgadari, C., Taub, D. D., Liao, F., Farber, J. M., Maheshwari, S., et al. (1995). Human interferon-inducible protein 10 is a potent inhibitor of angiogenesis in vivo. *Journal of Experimental Medicine, 182*(1), 155–162. https://doi.org/10.1084/jem.182.1.155.

Balkwill, F., Charles, K. A., & Mantovani, A. (2005). Smoldering and polarized inflammation in the initiation and promotion of malignant disease. *Cancer Cell,* 7(3), 211–217. https://doi.org/10.1016/j.ccr.2005.02.013.

Benítez-Bribiesca, L., Wong, A., Utrera, D., & Castellanos, E. (2001). The role of mast cell tryptase in neoangiogenesis of premalignant and malignant lesions of the uterine cervix. *Journal of Histochemistry and Cytochemistry, 49*(8), 1061–1062. https://doi.org/10.1177/002215540104900816.

Bennett, I. M., Zatsepina, O., Zamai, L., Azzoni, L., Mikheeva, T., & Perussia, B. (1996). Definition of a natural killer NKR-P1A+/CD56-/CD16- functionally immature human NK cell subset that differentiates in vitro in the presence of interleukin 12. *Journal of Experimental Medicine, 184*(5), 1845–1856. https://doi.org/10.1084/jem.184.5.1845.

Bingle, L., Brown, N. J., & Lewis, C. E. (2002). The role of tumour-associated macrophages in tumour progression: Implications for new anticancer therapies. *Journal of Pathology, 196*(3), 254–265. https://doi.org/10.1002/path.1027.

Blair, R. J., Meng, H., Marchese, M. J., Ren, S., Schwartz, L. B., Tonnesen, M. G., et al. (1997). Human mast cells stimulate vascular tube formation. Tryptase is a novel, potent angiogenic factor. *Journal of Clinical Investigation, 99*(11), 2691–2700. https://doi.org/10.1172/JCI119458.

Boesiger, J., Tsai, M., Maurer, M., Yamaguchi, M., Brown, L. F., Claffey, K. P., et al. (1998). Mast cells can secrete vascular permeability factor/vascular endothelial cell growth factor and exhibit enhanced release after immunoglobulin E- dependent upregulation of Fcε receptor I expression. *Journal of Experimental Medicine, 188*(6), 1135–1145. https://doi.org/10.1084/jem.188.6.1135.

Bourbié-Vaudaine, S., Blanchard, N., Hivroz, C., & Roméo, P. H. (2006). Dendritic cells can turn CD4+ T lymphocytes into vascular endothelial growth factor-carrying cells by intercellular neuropilin-1 transfer. *Journal of Immunology, 177*(3), 1460–1469. https://doi.org/10.4049/jimmunol.177.3.1460.

Bowrey, P. F., King, J., Magarey, C., Schwartz, P., Marr, P., Bolton, E., et al. (2000). Histamine, mast cells and tumour cell proliferation in breast cancer: Does preoperative cimetidine administration have an effect? *British Journal of Cancer, 82*(1), 167–170. https://doi.org/10.1054/bjoc.1999.0895.

Brown, L. F., Yeo, K. T., Berse, B., Yeo, T. K., Senger, D. R., Dvorak, H. F., et al. (1992). Expression of vascular permeability factor (vascular endothelial growth factor) by epidermal keratinocytes during wound healing. *Journal of Experimental Medicine, 176*(5), 1375–1379. https://doi.org/10.1084/jem.176.5.1375.

Bruno, A., Bassani, B., D'Urso, D. G., Pitaku, I., Cassinotti, E., Pelosi, G., et al. (2018). Angiogenin and the MMP9-TIMP2 axis are up-regulated in proangiogenic, decidual NK-like cells from patients with colorectal cancer. *The FASEB Journal, 32*(10), 5365–5377. https://doi.org/10.1096/fj.201701103r.

Bruno, A., Ferlazzo, G., & Albini, A. (2014). A think tank of TINK/TANKs: Tumor infiltrating/tumor-associated natural killer cells in tumor progression and angiogenesis. *Journal of the National Cancer Institute, 106*(8), dju200. https://doi.org/10.1093/jnci/dju200.

Bruno, A., Focaccetti, C., Pagani, A., Imperatori, A. S., Spagnoletti, M., Rotolo, N., et al. (2013). The proangiogenic phenotype of natural killer cells in patients with non-small cell lung cancer. *Neoplasia, 15*(2), 133–142. https://doi.org/10.1593/neo.121758.

Carrer, A., Moimas, S., Zacchigna, S., Pattarini, L., Zentilin, L., Ruozi, G., et al. (2012). Neuropilin-1 identifies a subset of bone marrow Gr1- monocytes that can induce tumor vessel normalization and inhibit tumor growth. *Cancer Research, 72*(24), 6371–6381. https://doi.org/10.1158/0008-5472.can-12-0762.

Carretero, R., Sektioglu, I. M., Garbi, N., Salgado, O. C., Beckhove, P., & Hämmerling, G. J. (2015). Eosinophils orchestrate cancer rejection by normalizing tumor vessels and enhancing infiltration of CD8 + T cells. *Nature Immunology, 16*(6), 609–617. https://doi.org/10.1038/ni.3159.

Cavallo, F., Quaglino, E., Cifaldi, L., Di Carlo, E., André, A., Bernabei, P., et al. (2001). Interleukin 12-activated lymphocytes influence tumor genetic programs. *Cancer Research, 61*(8), 3518–3523.
Chen, H., Campbell, R. A., Chang, Y., Li, M., Wang, C. S., Li, J., et al. (2009). Pleiotrophin produced by multiple myeloma induces transdifferentiation of monocytes into vascular endothelial cells: A novel mechanism of tumor-induced vasculogenesis. *Blood, 113*(9), 1992–2002. https://doi.org/10.1182/blood-2008-02-133751.
Chitu, V., & Stanley, E. R. (2006). Colony-stimulating factor-1 in immunity and inflammation. *Current Opinion in Immunology, 18*(1), 39–48. https://doi.org/10.1016/j.coi.2005.11.006.
Coffelt, S. B., Wellenstein, M. D., & De Visser, K. E. (2016). Neutrophils in cancer: Neutral no more. *Nature Reviews Cancer, 16*(7), 431–446. https://doi.org/10.1038/nrc.2016.52
Colombo, M. P., & Trinchieri, G. (2002). Interleukin-12 in anti-tumor immunity and immunotherapy. *Cytokine and Growth Factor Reviews, 13*(2), 155–168. https://doi.org/10.1016/S1359-6101(01)00032-6.
Conejo-Garcia, J. R., Benencia, F., Courreges, M.-C., Kang, E., Mohamed-Hadley, A., Buckanovich, R. J., et al. (2004). Tumor-infiltrating dendritic cell precursors recruited by a β-defensin contribute to vasculogenesis under the influence of Vegf-A. *Nature Medicine, 10*(9), 950–958. https://doi.org/10.1038/nm1097.
Cooper, M. A., Fehniger, T. A., Turner, S. C., Chen, K. S., Ghaheri, B. A., Ghayur, T., et al. (2001). Human natural killer cells: A unique innate immunoregulatory role for the CD56bright subset. *Blood, 97*(10), 3146–3151. https://doi.org/10.1182/blood.V97.10.3146.
Coussens, L. M., Raymond, W. W., Bergers, G., Laig-Webster, M., Behrendtsen, O., Werb, Z., et al. (1999). Inflammatory mast cells up-regulate angiogenesis during squamous epithelial carcinogenesis. *Genes and Development, 13*(11), 1382–1397. https://doi.org/10.1101/gad.13.11.1382.
Coussens, L. M., Tinkle, C. L., Hanahan, D., & Werb, Z. (2000). MMP-9 supplied by bone marrow-derived cells contributes to skin carcinogenesis. *Cell, 103*(3), 481–490. https://doi.org/10.1016/S0092-8674(00)00139-2.
Coussens, L. M., & Werb, Z. (1996). Matrix metalloproteinases and the development of cancer. *Chemistry and Biology, 3*(11), 895–904. https://doi.org/10.1016/S1074-5521(96)90178-7.
Coussens, L. M., & Werb, Z. (2002). Inflammation and cancer. *Nature, 420*(6917), 860–867. https://doi.org/10.1038/nature01322.
Curiel, T. J., Cheng, P., Mottram, P., Alvarez, X., Moons, L., Evdemon-Hogan, M., et al. (2004). Dendritic cell subsets differentially regulate angiogenesis in human ovarian cancer. *Cancer Research, 64*(16), 5535–5538. https://doi.org/10.1158/0008-5472.can-04-1272.
De Palma, M., & Jain, R. K. (2017). CD4+ T cell activation and vascular normalization: Two sides of the same coin? *Immunity, 46*(5), 773–775. https://doi.org/10.1016/j.immuni.2017.04.015.
De Visser, K. E., Korets, L. V., & Coussens, L. M. (2005). De novo carcinogenesis promoted by chronic inflammation is B lymphocyte dependent. *Cancer Cell,* 7(5), 411–423. https://doi.org/10.1016/j.ccr.2005.04.014.
DeNardo, D. G., Brennan, D. J., Rexhepaj, E., Ruffell, B., Shiao, S. L., Madden, S. F., et al. (2011). Leukocyte complexity predicts breast cancer survival and functionally regulates response to chemotherapy. *Cancer Discovery, 1*(1), 54–67. https://doi.org/10.1158/2159-8274.CD-10-0028.
Denhardt, D. T., Noda, M., O'Regan, A. W., Pavlin, D., & Berman, J. S. (2001). Osteopontin as a means to cope with environmental insults: Regulation of

inflammation, tissue remodeling, and cell survival. *Journal of Clinical Investigation, 107*(9), 1055–1061. https://doi.org/10.1172/JCI12980.

Deryugina, E. I., Zajac, E., Juncker-Jensen, A., Kupriyanova, T. A., Welter, L., & Quigley, J. P. (2014). Tissue-infiltrating neutrophils constitute the major in vivo source of angiogenesis-inducing MMP-9 in the tumor microenvironment. *Neoplasia, 16*(10), 771–788. https://doi.org/10.1016/j.neo.2014.08.013.

Detmar, M., Brown, L. F., Schön, M. P., Elicker, B. M., Velasco, P., Richard, L., et al. (1998). Increased microvascular density and enhanced leukocyte rolling and adhesion in the skin of VEGF transgenic mice. *Journal of Investigative Dermatology, 111*(1), 1–6. https://doi.org/10.1046/j.1523-1747.1998.00262.x.

Detoraki, A., Staiano, R. I., Granata, F., Giannattasio, G., Prevete, N., de Paulis, A., et al. (2009). Vascular endothelial growth factors synthesized by human lung mast cells exert angiogenic effects. *Journal of Allergy and Clinical Immunology, 123*(5), 1142–e5. https://doi.org/10.1016/j.jaci.2009.01.044.

Di Carlo, E., Forni, G., Lollini, P. L., Colombo, M. P., Modesti, A., & Musiani, P. (2001). The intriguing role of polymorphonuclear neutrophils in antitumor reactions. *Blood, 97*(2), 339–345. https://doi.org/10.1182/blood.V97.2.339.

Dias, S., Boyd, R., & Balkwill, F. (1998). IL-12 regulates VEGF and MMPs in a murine breast cancer model. *International Journal of Cancer, 78*(3), 361–365. https://doi.org/10.1002/(SICI)1097-0215(19981029)78:3<361::AID-IJC17>3.0.CO;2-9.

Dieu-Nosjean, M. C., Goc, J., Giraldo, N. A., Sautès-Fridman, C., & Fridman, W. H. (2014). Tertiary lymphoid structures in cancer and beyond. *Trends in Immunology, 35*(11), 571–580. https://doi.org/10.1016/j.it.2014.09.006.

Dunn, G. P., Bruce, A. T., Ikeda, H., Old, L. J., & Schreiber, R. D. (2002). Cancer immunoediting: From immunosurveillance to tumor escape. *Nature Immunology, 3*(11), 991–998. https://doi.org/10.1038/ni1102-991.

Dvorak, H. F. (1986). Tumors: Wounds that not heal. Similarities between tumor stroma generation and wound healing. *New England Journal of Medicine, 315*, 1650–1659.

Dvorak, H. (2006). Discovery of vascular permeability factor (VPF). *Experimental Cell Research, 312*(5), 522–526. https://doi.org/10.1016/j.yexcr.2005.11.026.

Dvorak, H. F. (2015). Tumors: Wounds that do not heal-redux. *Cancer Immunology Research, 3*(1), 1–11. https://doi.org/10.1158/2326-6066.CIR-14-0209.

Dvorak, H. F., Nagy, J. A., Berse, B., Brown, L. F., Yeo, K.-T., Yeo, T.-K., et al. (1992). Vascular permeability factor, fibrin, and the pathogenesis of tumor stroma formation. *Annals of the New York Academy of Sciences, 667*(1 Plasminogen A), 101–111. https://doi.org/10.1111/j.1749-6632.1992.tb51603.x.

Dvorak, H. F., Van Dewater, L., Dvorak, A. M., Harvey, V. S., Bitzer, A. M., Davis, G. L., et al. (1983). Procoagulant activity associated with plasma membrane vesicles shed by cultured tumor cells. *Cancer Research, 43*(9), 4434–4442.

Egami, K., Murohara, T., Shimada, T., Sasaki, K.-I., Shintani, S., Sugaya, T., et al. (2003). Role of host angiotensin II type 1 receptor in tumor angiogenesis and growth. *Journal of Clinical Investigation, 112*(1), 67–75. https://doi.org/10.1172/JCI16645.

Elpek, G. O., Gelen, T., Aksoy, N. H., Erdoğan, A., Dertsiz, L., Demircan, A., et al. (2001). The prognostic relevance of angiogenesis and mast cells in squamous cell carcinoma of the oesophagus. *Journal of Clinical Pathology, 54*(12), 940–944. https://doi.org/10.1136/jcp.54.12.940.

Engblom, C., Pfirschke, C., & Pittet, M. (2016). The role of myeloid cells in cancer therapies. *Nature Reviews Cancer, 6*(7), 447–462. https://doi.org/10.1038/nrc.2016.54.

Facciabene, A., Peng, X., Hagemann, I. S., Balint, K., Barchetti, A., Wang, L.-P., et al. (2011). Tumour hypoxia promotes tolerance and angiogenesis via CCL28 and Treg cells. *Nature, 475*(7355), 226–230. https://doi.org/10.1038/nature10169.

Finkin, S., Yuan, D., Stein, I., Taniguchi, K., Weber, A., Unger, K., et al. (2015). Ectopic lymphoid structures function as microniches for tumor progenitor cells in hepatocellular carcinoma. *Nature Immunology, 16*(12), 1235–1244. https://doi.org/10.1038/ni.3290.

Fitzsimons, C., Molinari, B., Duran, H., Palmieri, M., Davio, C., Cricco, G., et al. (1997). Atypical association of H1 and H2 histamine receptors with signal transduction pathways during multistage mouse skin carcinogenesis. *Inflammation Research, 46*(8), 292–298. https://doi.org/10.1007/s000110050190.

Freud, A. G., & Caligiuri, M. A. (2006). Human natural killer cell development. *Immunological Reviews, 214*(1), 56–72. https://doi.org/10.1111/j.1600-065X.2006.00451.x.

Fridlender, Z. G., Sun, J., Kim, S., Kapoor, V., Cheng, G., Ling, L., et al. (2009). Polarization of tumor-associated neutrophil phenotype by TGF-β: \N1\ versus \N2\ TAN. *Cancer Cell, 16*(3), 183–194. https://doi.org/10.1016/j.ccr.2009.06.017.

Fridman, W. H., Pagès, F., Sauts-Fridman, C., & Galon, J. (2012). The immune contexture in human tumours: Impact on clinical outcome. *Nature Reviews Cancer, 12*(4), 298–306. https://doi.org/10.1038/nrc3245.

Friedman, A. D. (2002). Transcriptional regulation of granulocyte and monocyte development. *Oncogene, 21*(21), 3377–3390. https://doi.org/10.1038/sj.onc.1205324.

Fukushima, N., Satoh, T., Sano, M., & Tokunaga, O. (2001). Angiogenesis and mast cells in non-Hodgkin's lymphoma: A strong correlation in angioimmunoblastic T-cell lymphoma. *Leukemia and Lymphoma, 42*(4), 709–720. https://doi.org/10.3109/10428190109099333.

Gabrilovich, D. I., Ostrand-Rosenberg, S., & Bronte, V. (2012). Coordinated regulation of myeloid cells by tumours. *Nature Reviews Immunology, 12*(4), 253–268. https://doi.org/10.1038/nri3175.

Gajewski, T. F., Schreiber, H., & Fu, Y. X. (2013). Innate and adaptive immune cells in the tumor microenvironment. *Nature Immunology, 14*(10), 1014–1022. https://doi.org/10.1038/ni.2703.

Gay, L. J., & Felding-Habermann, B. (2011). Contribution of platelets to tumour metastasis. *Nature Reviews Cancer, 11*(2), 123–134. https://doi.org/10.1038/nrc3004.

Glowacki, J., & Mulliken, J. B. (1982). Mast cells in hemangiomas and vascular malformations. *Pediatrics, 70*(1), 48–51.

Gobert, M., Treilleux, I., Bendriss-Vermare, N., Bachelot, T., Goddard-Leon, S., Arfi, V., et al. (2009). Regulatory T cells recruited through CCL22/CCR4 are selectively activated in lymphoid infiltrates surrounding primary breast tumors and lead to an adverse clinical outcome. *Cancer Research, 69*(5), 2000–2009. https://doi.org/10.1158/0008-5472.can-08-2360.

Gottfried, E., Kreutz, M., Haffner, S., Holler, E., Iacobelli, M., Andreesen, R., et al. (2007). Differentiation of human tumour-associated dendritic cells into endothelial-like cells: An alternative pathway of tumour angiogenesis. *Scandinavian Journal of Immunology, 65*(4), 329–335. https://doi.org/10.1111/j.1365-3083.2007.01903.x.

Gotthardt, D., Putz, E. M., Grundschober, E., Prchal-Murphy, M., Straka, E., Kudweis, P., et al. (2016). STAT5 is a key regulator in NK cells and acts as a molecular switch from tumor surveillance to tumor promotion. *Cancer Discovery, 6*(4), 414–429. https://doi.org/10.1158/2159-8290.cd-15-0732.

Graham, R. M., & Graham, J. B. (1966). Mast cells and cancer of the cervix. *Surgery Gynecology and Obstetrics, 123*(1), 3–9.

Grimbaldeston, M. A., Nakae, S., Kalesnikoff, J., Tsai, M., & Galli, S. J. (2007). Mast cell-derived interleukin 10 limits skin pathology in contact dermatitis and chronic irradiation with ultraviolet B. *Nature Immunology, 8*(10), 1095–1104. https://doi.org/10.1038/ni1503.

Gruber, B. L., Marchese, M. J., & Kew, R. (1995). Angiogenic factors stimulate mast-cell migration. *Blood, 86*(7), 2488–2493.

Grützkau, A., Krüger-Krasagakes, S., Baumeister, H., Schwarz, C., Kögel, H., Welker, P., et al. (1998). Synthesis, storage, and release of vascular endothelial growth factor/vascular permeability factor (VEGF/VPF) by human mast cells: Implications for the biological significance of VEGF206. *Molecular Biology of the Cell, 9*(4), 875–884. https://doi.org/10.1091/mbc.9.4.875.

Guiducci, C., Vicari, A. P., Sangaletti, S., Trinchieri, G., & Colombo, M. P. (2005). Redirecting in vivo elicited tumor infiltrating macrophages and dendritic cells towards tumor rejection. *Cancer Research, 65*(8), 3437–3446. https://doi.org/10.1158/0008-5472.CAN-04-4262.

Gutschalk, C. M., Herold-Mende, C. C., Fusenig, N. E., & Mueller, M. M. (2006). Granulocyte colony-stimulating factor and granulocyte-macrophage colony-stimulating factor promote malignant growth of cells from head and neck squamous cell carcinomas in vivo. *Cancer Research, 66*(16), 8026–8036. https://doi.org/10.1158/0008-5472.CAN-06-0158.

Hanada, T., Nakagawa, M., Emoto, A., Nomura, T., Nasu, N., & Nomura, Y. (2000). Prognostic value of tumor-associated macrophage count in human bladder cancer. *International Journal of Urology, 7*(7), 263–269. https://doi.org/10.1046/j.1442-2042.2000.00190.x.

Hartveit, F. (1981). Mast cells and metachromasia in human breast cancer: Their occurrence, significance and consequence: A preliminary report. *The Journal of Pathology, 134*(1), 7–11. https://doi.org/10.1002/path.1711340103.

Hayakawa, Y., & Smyth, M. J. (2006). Innate immune recognition and suppression of tumors. *Advances in Cancer Research, 95*, 293–322. https://doi.org/10.1016/S0065-230X(06)95008-8.

Hiraoka, N., Ino, Y., & Yamazaki-Itoh, R. (2016). Tertiary lymphoid organs in cancer tissues. Frontiers in Immunology, 7, 244–244. https://doi.org/10.3389/fimmu.2016.00244.

Hoeres, T., Wilhelm, M., Smetak, M., Holzmann, E., Schulze-Tanzil, G., & Birkmann, J. (2018). Immune cells regulate VEGF signalling via release of VEGF and antagonistic soluble VEGF receptor-1. *Clinical and Experimental Immunology, 192*(1), 54–67. https://doi.org/10.1111/cei.13090.

Huang, Y., Yuan, J., Righi, E., Kamoun, W. S., Ancukiewicz, M., Nezivar, J., et al. (2012). Vascular normalizing doses of antiangiogenic treatment reprogram the immunosuppressive tumor microenvironment and enhance immunotherapy. *Proceedings of the National Academy of Sciences of the United States of America, 109*(43), 17561–17566. https://doi.org/10.1073/pnas.1215397109.

Imada, A., Shijubo, N., Kojima, H., & Abe, S. (2000). Mast cells correlate angiogenesis and poor outcome in stage I lung adenocarcinoma. *European Respiratory Journal, 15*(6), 1087–1093. https://doi.org/10.1034/j.1399-3003.2000.01517.x.

Imtiyaz, H. Z., Williams, E. P., Hickey, M. M., Patel, S. A., Durham, A. C., Yuan, L. J., et al. (2010). Hypoxia-inducible factor 2α regulates macrophage function in mouse models of acute and tumor inflammation. *Journal of Clinical Investigation, 120*(8), 2699–2714. https://doi.org/10.1172/JCI39506.

Ishibashi, S., Ohashi, Y., Suzuki, T., Miyazaki, S., Moriya, T., Satomi, S., et al. (2006). Tumor-associated tissue eosinophilia in human esophageal squamous cell carcinoma. *Anticancer Research, 26*(2B), 1419–1424. https://www.ncbi.nlm.nih.gov/pubmed/16619553.

Ivanova, E. A., & Orekhov, A. N. (2015). T helper lymphocyte subsets and plasticity in autoimmunity and cancer: an overview. *Biomedical Research International, 327470.* https://doi.org/10.1155/2015/327470.

Jenkins, D. C., Charles, I. G., Thomsen, L. L., Moss, D. W., Holmes, L. S., Baylis, S. A., et al. (1995). Roles of nitric oxide in tumor growth. *Proceedings of the National Academy of*

Sciences of the United States of America, 92(10), 4392–4396. https://doi.org/10.1073/pnas.92.10.4392.

Jeong, H.-J., Oh, H.-A., Nam, S.-Y., Han, N.-R., Kim, Y.-S., Kim, J.-H., et al. (2012). The critical role of mast cell-derived hypoxia-inducible factor-1α in human and mice melanoma growth. *International Journal of Cancer, 132*(11), 2492–2501. https://doi.org/10.1002/ijc.27937.

Jones, D., Pereira, E. R., & Padera, T. P. (2018). Growth and immune evasion of lymph node metastasis. *Frontiers in Oncology, 8*. https://doi.org/10.3389/fonc.2018.00036.

Kanbe, N., Kurosawa, M., Nagata, H., Yamashita, T., Kurimoto, F., & Miyachi, Y. (2000). Production of fibrogenic cytokines by cord blood-derived cultured human mast cells. *Journal of Allergy and Clinical Immunology, 106*(1), S85–S90. https://doi.org/10.1067/mai.2000.106777.

Kinet, J. P. (2007). The essential role of mast cells in orchestrating inflammation. *Immunological Reviews, 217*(1), 5–7. https://doi.org/10.1111/j.1600-065X.2007.00528.x.

Klement, G. L., Yip, T.-T., Cassiola, F., Kikuchi, L., Cervi, D., Podust, V., et al. (2009). Platelets actively sequester angiogenesis regulators. *Blood, 113*(12), 2835–2842. https://doi.org/10.1182/blood-2008-06-159541.

Klimp, A. H., Hollema, H., Kempinga, C., van der Zee, A. G. J., de Vries, E. G. E., & Daemen, T. (2001). Expression of cyclooxygenase-2 and inducible nitric oxide synthase in human ovarian tumors and tumor-associated macrophages. *Cancer Research, 61*(19), 7305–7309.

Koehne, C. H., & Dubois, R. N. (2004). COX-2 inhibition and colorectal cancer. *Seminars in Oncology, 31*(7), 12–21. https://doi.org/10.1016/j.seminoncol.2004.03.041.

Koide, N., Nishio, A., Sato, T., Sugiyama, A., & Miyagawa, S. I. (2004). Significance of macrophage chemoattractant protein-1 expression and macrophage infiltration in squamous cell carcinoma of the esophagus. *American Journal of Gastroenterology, 99*(9), 1667–1674. https://doi.org/10.1111/j.1572-0241.2004.30733.x.

Kominsky, D. J., Campbell, E. L., & Colgan, S. P. (2010). Metabolic shifts in immunity and inflammation. *Journal of Immunology, 184*(8), 4062–4068. https://doi.org/10.4049/jimmunol.0903002.

Koskivirta, I., Rahkonen, O., Mäyränpää, M., Pakkanen, S., Husheem, M., Sainio, A., et al. (2006). Tissue inhibitor of metalloproteinases 4 (TIMP4) is involved in inflammatory processes of human cardiovascular pathology. *Histochemistry and Cell Biology, 126*(3), 335–342. https://doi.org/10.1007/s00418-006-0163-8.

Krzywinska, E., Kantari-Mimoun, C., Kerdiles, Y., Sobecki, M., Isagawa, T., Gotthardt, D., et al. (2017). Loss of HIF-1α in natural killer cells inhibits tumour growth by stimulating non-productive angiogenesis. *Nature Communications, 8*(1), 1597. https://doi.org/10.1038/s41467-017-01599-w.

Labelle, M., Begum, S., & Hynes, R. O. (2014). Platelets guide the formation of early metastatic niches. *Proceedings of the National Academy of Sciences of the United States of America, 111*(30), E3053–E3061. https://doi.org/10.1073/pnas.1411082111.

Lachter, J., Stein, M., Lichtig, C., Eidelman, S., & Munichor, M. (1995). Mast cells in colorectal neoplasias and premalignant disorders. *Diseases of the Colon and Rectum, 38*(3), 290–293. https://doi.org/10.1007/BF02055605.

Leali, D., Dell'Era, P., Stabile, H., Sennino, B., Chambers, A. F., Naldini, A., et al. (2003). Osteopontin (Eta-1) and fibroblast growth factor-2 cross-talk in angiogenesis. *Journal of Immunology, 171*(2), 1085–1093. https://doi.org/10.4049/jimmunol.171.2.1085.

Leek, R. D., Landers, R. J., Harris, A. L., & Lewis, C. E. (1999). Necrosis correlates with high vascular density and focal macrophage infiltration in invasive carcinoma of the breast. *British Journal of Cancer, 79*(5–6), 991–995. https://doi.org/10.1038/sj.bjc.6690158.

Lee, H. S., Schlereth, S. L., Park, E. Y., Emami-Naeini, P., Chauhan, S. K., & Dana, R. (2014). A novel pro-angiogenic function for interferon-γ-secreting natural killer cells. *Investigative Ophthalmology and Visual Science, 55*(5), 2885–2892. https://doi.org/10.1167/iovs.14-14093.

Lin, E. Y., Li, J. F., Bricard, G., Wang, W., Deng, Y., Sellers, R., et al. (2007). Vascular endothelial growth factor restores delayed tumor progression in tumors depleted of macrophages. *Molecular Oncology, 1*(3), 288–302. https://doi.org/10.1016/j.molonc.2007.10.003.

Lin, E. Y., Li, J. F., Gnatovskiy, L., Deng, Y., Zhu, L., Grzesik, D. A., et al. (2006). Macrophages regulate the angiogenic switch in a mouse model of breast cancer. *Cancer Research, 66*(23), 11238–11246. https://doi.org/10.1158/0008-5472.CAN-06-1278.

Lin, E. Y., Nguyen, A. V., Russell, R. G., & Pollard, J. W. (2001). Colony-stimulating factor 1 promotes progression of mammary tumors to malignancy. *Journal of Experimental Medicine, 193*(6), 727–739. https://doi.org/10.1084/jem.193.6.727.

Lissbrant, I. F., Stattin, P., Wikstrom, P., Damber, J. E., Egevad, L., & Bergh, A. (2000). Tumor associated macrophages in human prostate cancer: Relation to clinicopathological variables and survival. *International Journal of Oncology, 17*(3), 445–451.

Liu, C., Workman, C. J., & Vignali, D. A. A. (2016). Targeting regulatory T cells in tumors. *FEBS Journal*, 2731–2748. https://doi.org/10.1111/febs.13656.

Mantovani, A., Sozzani, S., Locati, M., Allavena, P., & Sica, A. (2002). Macrophage polarization: Tumor-associated macrophages as a paradigm for polarized M2 mononuclear phagocytes. *Trends in Immunology, 23*(11), 549–555. https://doi.org/10.1016/S1471-4906(02)02302-5.

Marinaccio, C., Ingravallo, G., Gaudio, F., Perrone, T., Nico, B., Maoirano, E., et al. (2014). Microvascular density, CD68 and tryptase expression in human diffuse large B-cell lymphoma. *Leukemia Research, 38*(11), 1374–1377. https://doi.org/10.1016/j.leukres.2014.09.007.

Marinaccio, C., Ingravallo, G., Gaudio, F., Perrone, T., Ruggieri, S., Opinto, G., et al. (2016). T cells, mast cells and microvascular density in diffuse large B cell lymphoma. *Clinical and Experimental Medicine, 16*(3), 301–306. https://doi.org/10.1007/s10238-015-0354-5.

de Martel, C., Ferlay, J., Franceschi, S., Vignat, J., Bray, F., Forman, D., et al. (2012). Global burden of cancers attributable to infections in 2008: A review and synthetic analysis. *The Lancet Oncology, 13*(6), 607–615. https://doi.org/10.1016/s1470-2045(12)70137-7.

Masson, V., de la Ballina, L. R., Munaut, C., Wielockx, B., Jost, M., Maillard, C., et al. (2005). Contribution of host MMP-2 and MMP-9 to promote tumor vascularization and invasion of malignant keratinocytes. *Federation of American Societies for Experimental Biology Journal: Official Publication of the Federation of American Societies for Experimental Biology, 19*(2), 234–236. https://doi.org/10.1096/fj.04-2140fje.

Metcalfe, D. D., Baram, D., & Mekori, Y. A. (1997). Mast cells. *Physiological Reviews,* 77(4), 1033–1079. https://doi.org/10.1152/physrev.1997.77.4.1033.

Miller, J. S., Verfaillie, C., & McGlave, P. (1992). The generation of human natural killer cells from CD34+/DR- primitive progenitors in long-term bone marrow culture. *Blood, 80*(9), 2182–2187.

Mitchell, D. A., Batich, K. A., Gunn, M. D., Huang, M.-N., Sanchez-Perez, L., Nair, S. K., et al. (2015). Tetanus toxoid and CCL3 improve dendritic cell vaccines in mice and glioblastoma patients. *Nature, 519*(7543), 366–369. https://doi.org/10.1038/nature14320.

Molica, S., Vacca, A., Crivellato, E., Cuneo, A., & Ribatti, D. (2003). Tryptase-positive mast cells predict clinical outcome of patients with early B-cell chronic lymphocytic leukemia. *European Journal of Haematology, 71*(2), 137–139. https://doi.org/10.1034/j.1600-0609.2003.00110.x.

Moller, A., Lippert, U., Lessmann, D., Kolde, G., Hamann, K., Welker, P., et al. (1993). Human mast cells produce IL-8. *Journal of Immunology, 151*(6), 3261–3266.

Moretta, A., Bottino, C., Vitale, M., Pende, D., Cantoni, C., Cristina, M., et al. (2001). Activating receptors and coreceptors involved in human natural killer cell-mediated cytolysis. *Annual Review of Immunology, 19*, 197–223. https://doi.org/10.1146/annurev.immunol.19.1.197.

Movahedi, K., Laoui, D., Gysemans, C., Baeten, M., Stange, G., Van den Bossche, J., et al. (2010). Different tumor microenvironments contain functionally distinct subsets of macrophages derived from Ly6C(high) monocytes. *Cancer Research, 70*(14), 5728–5739. https://doi.org/10.1158/0008-5472.can-09-4672.

Munitz, A., & Levi-Schaffer, F. (2004). Eosinophils: "new" roles for "old" cells. *Allergy, 59*(3), 268–275. https://doi.org/10.1111/j.1398-9995.2003.00442.x.

Nagy, J. A., Benjamin, L., Zeng, H., Dvorak, A. M., & Dvorak, H. F. (2008). Vascular permeability, vascular hyperpermeability and angiogenesis. *Angiogenesis, 11*(2), 109–119.

Nakao, S., Kuwano, T., Tsutsumi-Miyahara, C., Ueda, S., Kimura, Y. N., Hamano, S., et al. (2005). Infiltration of COX-2-expressing macrophages is a prerequisite for IL-1 beta-induced neovascularization and tumor growth. *Journal of Clinical Investigation, 115*(11), 2979–2991. https://doi.org/10.1172/JCI23298.

Naldini, A., & Carraro, F. (2005). Role of inflammatory mediators in angiogenesis. *Current Drug Targets Inflammation and Allergy, 4*(1), 3–8. https://doi.org/10.2174/1568010053622830.

Naldini, A., Leali, D., Pucci, A., Morena, E., Carraro, F., Nico, B., et al. (2006). Cutting edge: IL-1β mediates the proangiogenic activity of osteopontin-activated human monocytes. *Journal of Immunology, 177*(7), 4267–4270.

Nico, B., Mangieri, D., Crivellato, E., Vacca, A., & Ribatti, D. (2008). Mast cells contribute to vasculogenic mimicry in multiple myeloma. *Stem Cells and Development, 17*(1), 19–22. https://doi.org/10.1089/scd.2007.0132.

Nielsen, H. J., Hansen, U., Christensen, I. J., Reimert, C. M., Brünner, N., & Moesgaard, F. (1999). Independent prognostic value of eosinophil and mast cell infiltration in colorectal cancer tissue. *Journal of Pathology, 189*(4), 487–495. https://doi.org/10.1002/(SICI)1096-9896(199912)189:4<487::AID-PATH484>3.0.CO;2-I.

Nilsson, G., Forsberg-Nilsson, K., Xiang, Z., Hallböök, F., Nilsson, K., & Metcalfe, D. D. (1997). Human mast cells express functional TrkA and are a source of nerve growth factor. *European Journal of Immunology, 27*(9), 2295–2301. https://doi.org/10.1002/eji.1830270925.

Nishie, A., Ono, M., Shono, T., Fukushi, J., Otsubo, M., Onoue, H., et al. (1999). Macrophage infiltration and heme oxygenase-1 expression correlate with angiogenesis in human gliomas. *Clinical Cancer Research, 5*(5), 1107–1113. https://www.ncbi.nlm.nih.gov/pubmed/10353745.

Norrby, K., Jakobsson, A., & Sörbo, J. (1986). Mast-cell-mediated angiogenesis: A novel experimental model using the rat mesentery. *Virchows Archiv B Cell Pathology Including Molecular Pathology, 52*(1), 195–206. https://doi.org/10.1007/BF02889963.

Norrby, K., Jakobsson, A., & Sörbo, J. (1989). Mast cell secretion and angiogenesis, a quantitative study in rats and mice. *Virchows Archiv B Cell Pathology Including Molecular Pathology, 57*, 251–257.

Novitskiy, S. V., Ryzhov, S., Zaynagetdinov, R., Goldstein, A. E., Huang, Y., Tikhomirov, O. Y., et al. (2008). Adenosine receptors in regulation of dendritic cell differentiation and function. *Blood, 112*(5), 1822–1831. https://doi.org/10.1182/blood-2008-02-136325.

Nozawa, H., Chiu, C., & Hanahan, D. (2006). Infiltrating neutrophils mediate the initial angiogenic switch in a mouse model of multistage carcinogenesis. *Proceedings of the National Academy of Sciences of the United States of America, 103*(33), 12493–12498. https://doi.org/10.1073/pnas.0601807103.

Obermueller, E., Vosseler, S., Fusenig, N. E., & Mueller, M. M. (2004). Cooperative autocrine and paracrine functions of granulocyte colony-stimulating factor and granulocyte-macrophage colony-stimulating factor in the progression of skin carcinoma cells. *Cancer Research, 64*(21), 7801–7812. https://doi.org/10.1158/0008-5472.CAN-03-3301.

Ohmura, Y., Yoshikawa, K., Saga, S., Ueda, R., Kazaoka, Y., & Yamada, S. (2008). Combinations of tumor-specific CD8+ CTLs and anti-CD25 mAb provide improved immunotherapy. *Oncology Reports, 19*(5), 1265–1270.

Ohno, S., Ohno, Y., Suzuki, N., Kamei, T., Koike, K., Inagawa, H., et al. (2004). Correlation of histological localization of tumor-associated macrophages with clinicopathological features in endometrial cancer. *Anticancer Research, 24*(Issue 5 C), 3335–3342.

Ohno, I., Ohtani, H., Nitta, Y., Suzuki, J., Hoshi, H., Honma, M., et al. (1997). Eosinophils as a source of matrix metalloproteinase-9 in asthmatic airway inflammation. *American Journal of Respiratory Cell and Molecular Biology, 16*(3), 212–219. https://doi.org/10.1165/ajrcmb.16.3.9070604.

Ostroumov, D., Fekete-Drimusz, N., Saborowski, M., Kühnel, F., & Woller, N. (2018). CD4 and CD8 T lymphocyte interplay in controlling tumor growth. *Cellular and Molecular Life Sciences, 75*(4), 689–713. https://doi.org/10.1007/s00018-017-2686-7.

Palazon, A., Tyrakis, P. A., Macias, D., Veliça, P., Rundqvist, H., Fitzpatrick, S., et al. (2017). An HIF-1α/VEGF-A Axis in cytotoxic T cells regulates tumor progression. *Cancer Cell, 32*(5), 669–683.e5. https://doi.org/10.1016/j.ccell.2017.10.003.

Palumbo, J. S., Talmage, K. E., Massari, J. V., La Jeunesse, C. M., Flick, M. J., Kombrinck, K. W., et al. (2005). Platelets and fibrin(ogen) increase metastatic potential by impeding natural killer cell-mediated elimination of tumor cells. *Blood, 105*(1), 178–185. https://doi.org/10.1182/blood-2004-06-2272.

Park, J., Wysocki, R. W., Amoozgar, Z., Maiorino, L., Fein, M. R., Jorns, J., et al. (2016). Cancer cells induce metastasis-supporting neutrophil extracellular DNA traps. *Science Translational Medicine, 8*(361), 361ra138. https://doi.org/10.1126/scitranslmed.aag1711.

Perdiguero, E. G., & Geissmann, F. (2016). The development and maintenance of resident macrophages. *Nature Immunology, 17*(1), 2–8. https://doi.org/10.1038/ni.3341.

Pittoni, P., Tripodo, C., Piconese, S., Mauri, G., Parenza, M., Rigoni, A., et al. (2011). Mast cell targeting hampers prostate adenocarcinoma development but promotes the occurrence of highly malignant neuroendocrine cancers. *Cancer Research, 71*(18), 5987–5997. https://doi.org/10.1158/0008-5472.CAN-11-1637.

Puxeddu, I., Alian, A., Piliponsky, A. M., Ribatti, D., Panet, A., & Levi-Schaffer, F. (2005). Human peripheral blood eosinophils induce angiogenesis. *The International Journal of Biochemistry and Cell Biology, 37*(3), 628–636. https://doi.org/10.1016/j.biocel.2004.09.001.

Qian, B., & Pollard, J. M. (2010). Macrophage density enhances tumor progression and metastasis. *Cell, 141*, 39–45.

Queen, M. M., Ryan, R. E., Holzer, R. G., Keller-Peck, C. R., & Jorcyk, C. L. (2005). Breast cancer cells stimulate neutrophils to produce oncostatin M: Potential implications for tumor progression. *Cancer Research, 65*(19), 8896–8904. https://doi.org/10.1158/0008-5472.can-05-1734.

Qu, Z., Huang, X., Ahmadi, P., Stenberg, P., Liebler, J. M., Le, A. C., et al. (1998). Synthesis of basic fibroblast growth factor by murine mast cells. Regulation by transforming growth factor beta, tumor necrosis factor alpha, and stem cell factor. *International Archives of Allergy and Immunology, 115*(1), 47–54. https://doi.org/10.1159/000023829.

Qu, Z., Kayton, & Ahmadi, P. (1998). Chemopreventive properties of a selective inducible nitric oxide synthase inhibitor in colon carcinogenesis, administered alone or in combination with celecoxib, a selective cyclooxygenase-2 inhibitor. *Journal of Histochemistry and Cytochemistry, 46*, 165–170.

Rao, C. V., Indranie, C., Simi, B., Reddy, B. S., Manning, P. T., & Connor, J. R. (2002). Chemopreventive properties of a selective inducible nitric oxide synthase inhibitor in colon carcinogenesis, administered alone or in combination with celecoxib, a selective cyclooxygenase-2 inhibitor. *Cancer Research, 62*(1), 165–170.

Reed, J. A., Scott McNutt, N., Bogdany, J. K., & Albino, A. P. (1996). Expression of the mast cell growth factor interleukin-3 in melanocytic lesions correlates with an increased number of mast cells in the perilesional stroma: Implications for melanoma progression. *Journal of Cutaneous Pathology, 23*(6), 495–505. https://doi.org/10.1111/j.1600-0560.1996.tb01441.x.

Ribatti, D., & Crivellato, E. (2009). The controversial role of mast cells in tumor growth. *International Review of Cell and Molecular Biology, 275*, 89–131. https://doi.org/10.1016/S1937-6448(09)75004-X.

Ribatti, D., Crivellato, E., Candussio, L., Nico, B., Vacca, A., Roncali, L., et al. (2001). Mast cells and their secretory granules are angiogenic in the chick embryo chorioallantoic membrane. *Clinical and Experimental Allergy, 31*(4), 602–608. https://doi.org/10.1046/j.1365-2222.2001.00986.x.

Ribatti, D., Nico, B., Vacca, A., Marzullo, A., Calvi, N., Roncali, L., et al. (1998). Do mast cells help to induce angiogenesis in B-cell non-Hodgkin's lymphomas? *British Journal of Cancer,* 77(11), 1900–1906. https://doi.org/10.1038/bjc.1998.316.

Ribatti, D., Polimeno, G., Vacca, A., Marzullo, A., Crivellato, E., Nico, B., et al. (2002). Correlation of bone marrow angiogenesis and mast cells with tryptase activity in myelodysplastic syndromes. *Leukemia, 16*(9), 1680–1684. https://doi.org/10.1038/sj.leu.2402586.

Ribatti, D., Roncali, L., Nico, B., & Bertossi, M. (1987). Effects of exogenous heparin on the vasculogenesis of the chorioallantoic membrane. *Cells Tissues Organs, 130*(3), 257–263. https://doi.org/10.1159/000146454.

Ribatti, D., Vacca, A., Nico, B., Quondamatteo, F., Ria, R., Minischetti, M., et al. (1999). Bone marrow angiogenesis and mast cell density increase simultaneously with progression of human multiple myeloma. *British Journal of Cancer, 79*(3–4), 451–455. https://doi.org/10.1038/sj.bjc.6690070.

Ribatti, D., Vacca, A., & Ria, R. (2003). Neovascularization, expression of basic fibroblast growth factor-2, and mast cells with tryptase activity increase simultaneously with pathological progression in human malignant melanoma. *European Journal of Cancer, 39*, 666–675.

Riboldi, E., Musso, T., Moroni, E., Urbinati, C., Bernasconi, S., Rusnati, M., et al. (2005). Cutting edge: Proangiogenic properties of alternatively activated dendritic cells. *Journal of Immunology, 175*(5), 2788–2792. https://doi.org/10.4049/jimmunol.175.5.2788.

Rolny, C., Mazzone, M., Tugues, S., Laoui, D., Johansson, I., Coulon, C., et al. (2011). HRG inhibits tumor growth and metastasis by inducing macrophage polarization and vessel normalization through downregulation of PlGF. *Cancer Cell, 19*(1), 31–44. https://doi.org/10.1016/j.ccr.2010.11.009.

Romagnani, P., Annunziato, F., Lasagni, L., Lazzeri, E., Beltrame, C., Francalanci, M., et al. (2001). Cell cycle-dependent expression of CXC chemokine receptor 3 by endothelial cells mediates angiostatic activity. *The Journal of Clinical Investigation, 107*(1), 53–63. https://doi.org/10.1172/JCI9775.

Rothwell, P. M., Wilson, M., Price, J. F., Belch, J. F. F., Meade, T. W., & Mehta, Z. (2012). Effect of daily aspirin on risk of cancer metastasis: A study of incident cancers during randomised controlled trials. *The Lancet, 379*(9826), 1591–1601. https://doi.org/10.1016/S0140-6736(12)60209-8.

Salcedo, R., Young, H. A., Ponce, M. L., Ward, J. M., Kleinman, H. K., Murphy, W. J., et al. (2001). Eotaxin (CCL11) induces in vivo angiogenic responses by human CCR3+ endothelial cells. *Journal of Immunology, 166*(12), 7571–7578. https://doi.org/10.4049/jimmunol.166.12.7571.

Sawatsubashi, M., Yamada, T., Fukushima, N., Mizokami, H., Tokunaga, O., & Shin, T. (2000). Association of vascular endothelial growth factor and mast cells with angiogenesis in laryngeal squamous cell carcinoma. *Virchows Archiv, 436*(3), 243–248. https://doi.org/10.1007/s004280050037.

Scapini, P., Lapinet-Vera, J. A., Gasperini, S., Calzetti, F., Bazzoni, F., & Cassatella, M. A. (2000). The neutrophil as a cellular source of chemokines. *Immunological Reviews, 177*, 195–203. https://doi.org/10.1034/j.1600-065X.2000.17706.x.

Scavelli, C., Nico, B., Cirulli, T., Ria, R., Di Pietro, G., Mangieri, D., et al. (2007). Vasculogenic mimicry by bone marrow macrophages in patients with multiple myeloma. *Oncogene, 27*(5), 663–674. https://doi.org/10.1038/sj.onc.1210691.

Schmileau, J., & Finn, O. (2011). Activated granulocytes and granulocyte-derived hydrogen peroxide are the underlying mechanism of suppression of T cell function in advanced cancer patients. *Cancer Research, 61*, 4756–4760.

Schreiber, R. D., Old, L. J., & Smyth, M. J. (2011). Cancer immunoediting: Integrating immunity's roles in cancer suppression and promotion. *Science, 331*(6024), 1565–1570. https://doi.org/10.1126/science.1203486.

Shankaran, V., Ikeda, H., Bruce, A. T., White, J. M., Swanson, P. E., Old, L. J., et al. (2001). IFNγ, and lymphocytes prevent primary tumour development and shape tumour immunogenicity. *Nature, 410*(6832), 1107–1111. https://doi.org/10.1038/35074122.

Shojaei, F., Singh, M., Thompson, J. D., & Ferrara, N. (2008). Role of Bv8 in neutrophil-dependent angiogenesis in a transgenic model of cancer progression. *Proceedings of the National Academy of Sciences of the United States of America, 105*(7), 2640–2645. https://doi.org/10.1073/pnas.0712185105.

Shojaei, F., Wu, X., Malik, A. K., Zhong, C., Baldwin, M. E., Schanz, S., et al. (2007). Tumor refractoriness to anti-VEGF treatment is mediated by CD11b +Gr1+ myeloid cells. *Nature Biotechnology, 25*(8), 911–920. https://doi.org/10.1038/nbt1323.

Shurin, G. V., Ma, Y., & Shurin, M. R. (2013). Immunosuppressive mechanisms of regulatory dendritic cells in cancer. *Cancer Microenvironment, 6*(2), 159–167. https://doi.org/10.1007/s12307-013-0133-3.

Sica, A., Schioppa, T., Mantovani, A., & Allavena, P. (2006). Tumour-associated macrophages are a distinct M2 polarised population promoting tumour progression: Potential targets of anti-cancer therapy. *European Journal of Cancer, 42*(6), 717–727. https://doi.org/10.1016/j.ejca.2006.01.003.

Skobe, M., Hamberg, L. M., Hawighorst, T., Schirner, M., Wolf, G. L., Alitalo, K., et al. (2001). Concurrent induction of lymphangiogenesis, angiogenesis, and macrophage recruitment by vascular endothelial growth factor-C in melanoma. *American Journal of Pathology, 159*(3), 893–903. https://doi.org/10.1016/S0002-9440(10)61765-8.

Smyth, M. J., Hayakawa, Y., Takeda, K., & Yagita, H. (2002). New aspects of natural-killer-cell surveillance and therapy of cancer. *Nature Reviews Cancer, 2*(11), 850–861. https://doi.org/10.1038/nrc928.

Sörbo, J., Jakobbson, A., & Norrby, K. (1994). Mast cell histamine is angiogenic through receptor sfor histamine 1 and histamine 2. *International Journal of Experimental Pathology, 75*, 43–50.

de Souza, D. A., Jr., Toso, V. D., Campos, M. R. de C., Lara, V. S., Oliver, C., & Jamur, M. C. (2012). Expression of mast cell proteases correlates with mast cell maturation and angiogenesis during tumor progression. *PLoS One,* 7(7), e40790. https://doi.org/10.1371/journal.pone.0040790.

Starkey, J. R., Crowle, P. K., & Taubenberger, S. (1988). Mast-cell-deficient W/Wv mice exhibit A decreased rate of tumor angiogenesis. *International Journal of Cancer, 42*(1), 48–52. https://doi.org/10.1002/ijc.2910420110.

Stockmann, C., Doedens, A., Weidemann, A., Zhang, N., Takeda, N., Greenberg, J. I., et al. (2008). Deletion of vascular endothelial growth factor in myeloid cells accelerates tumorigenesis. *Nature, 456*(7223), 814–819. https://doi.org/10.1038/nature07445.

Sullivan, J. L., & Woda, B. A. (1989). X-linked lymphoproliferative syndrome. *Immunodeficiency Reviews, 1*(4), 325–347.

Takanami, I., Takeuchi, K., & Naruke, M. (2000). Mast cell density is associated with angiogenesis and poor prognosis in pulmonary adenocarcinoma. *Cancer, 88*(12), 2686–2692. https://doi.org/10.1002/1097-0142(20000615)88:12<2686::AID-CNCR6>3.0.CO;2-6.

Talmadge, J. E., & Gabrilovich, D. I. (2013). History of myeloid-derived suppressor cells. *Nature Reviews Cancer, 13*(10), 739–752. https://doi.org/10.1038/nrc3581.

Tanaka, A., Yamane, Y., & Matsuda, H. (2001). Mast cell MMP-9 production enhanced by bacterial lipopolysaccharide. *Journal of Veterinary Medical Science, 63*(7), 811–813. https://doi.org/10.1292/jvms.63.811.

Tazzyman, S., Lewis, C. E., & Murdoch, C. (2009). Neutrophils: Key mediators of tumour angiogenesis. *International Journal of Experimental Pathology, 90*(3), 222–231. https://doi.org/10.1111/j.1365-2613.2009.00641.x.

Tomita, M., Matsuzaki, Y., & Onitsuka, T. (2000). Effect of mast cells on tumor angiogenesis in lung cancer. *Annals of Thoracic Surgery, 69*(6), 1686–1690. https://doi.org/10.1016/S0003-4975(00)01160-7.

Trinchieri, G. (1989). Biology of natural killer cells. *Advances in Immunology, 47*(C), 187–376. https://doi.org/10.1016/S0065-2776(08)60664-1.

Trinchieri, G. (1993). Interleukin-12 and its role in the generation of TH1 cells. *Immunology Today, 14*(7), 335–338. https://doi.org/10.1016/0167-5699(93)90230-I.

Ullah, E., Nagi, A. H., & Lail, R. (2012). Angiogenesis and mast cell density in invasive pulmonary adenocarcinoma. *Journal of Cancer Research and Therapeutics, 8*(4), 537–541. https://doi.org/10.4103/0973-1482.106530.

Ullrich, S. E., Nghiem, D. X., & Khaskina, P. (2007). Suppression of an established immune response by UVA - a critical role for mast cells. *Photochemistry and Photobiology, 83*(5), 1095–1100. https://doi.org/10.1111/j.1751-1097.2007.00184.x.

Vacca, P., Moretta, L., Moretta, A., & Mingari, M. C. (2011). Origin, phenotype and function of human natural killer cells in pregnancy. *Trends in Immunology, 32*(11), 517–523. https://doi.org/10.1016/j.it.2011.06.013.

Walsh, L. J., Trinchieri, G., Waldorf, H. A., Whitaker, D., & Murphy, G. F. (1991). Human dermal mast cells contain and release tumor necrosis factor α, which induces endothelial leukocyte adhesion molecule 1. *Proceedings of the National Academy of Sciences of the United States of America, 88*(10), 4220–4224. https://doi.org/10.1073/pnas.88.10.4220.

Wong, S. Y., & Reiter, J. F. (2011). Wounding mobilizes hair follicle stem cells to form tumors. *Proceedings of the National Academy of Sciences of the United States of America, 108*(10), 4093–4098. https://doi.org/10.1073/pnas.1013098108.

Yano, H., Kinuta, M., Tateishi, H., Nakano, Y., Matsui, S., Monden, T., et al. (1999). Mast cell infiltration around gastric cancer cells correlates with tumor angiogenesis and metastasis. *Gastric Cancer, 2*(1), 26–32. https://doi.org/10.1007/s101200050017.

Yao, L., Sgadari, C., Furuke, K., Bloom, E. T., Teruya-Feldstein, J., & Tosato, G. (1999). Contribution of natural killer cells to inhibition of angiogenesis by interleukin-12. *Blood, 93*(5), 1612–1621.

Zhang, Q.w., Liu, L., Gong, C.y., Shi, H.s., Zeng, Y.h., Wang, X.z., et al. (2012). Prognostic significance of tumor-associated macrophages in solid tumor: A meta-analysis of the literature. *PLoS One,* 7(12). https://doi.org/10.1371/journal.pone.0050946.

CHAPTER 10

Other cells of the tumor microenvironment

10.1 Tie-2 expressing monocyte

Tie-2 expressing monocytes (TEMs) were not observed in normal tissues suggesting that they may represent a specific subset of resident monocytes (Venneri et al., 2007). The Tie-2 tyrosine kinase receptor is expressed only by endothelial cells and hematopoietic stem cells (HSCs) (Jones et al., 2001). Whereas only 1−2 per cent of total leukocytes are Tie-2^+, a substantial fraction (~20 per cent) of circulating monocytes express Tie-2 (Venneri et al., 2007). The majority of the circulating TEM do not express endothelial cell/endothelial precursor cell (EPC) markers, such as vascular endothelial growth factor receptor-2 (VEGFR-2), AC133, CD146, and CD34, whereas they express hematopoietic markers, such as CD45. Moreover, circulating human TEMs do not express C−C chemokine receptor-2 (CCR-2), the receptor for monocyte chemoattractant protein-1 (MCP-1), a chemokine that regulates the recruitment of monocytes to inflamed tissues and tumors. TEMs might be attracted to tumors in a CCR-2 independent manner, by signals produced by tumor cells, stromal cells, or endothelial cells. TEMs are recruited to tumors by C-X-C Motif Chemokine Ligand 12(CXCL12) via their expression of chemokine receptor CXCR4 (Welford et al., 2011). Indeed, Tie-2^+ CD14^+ cells isolated from human peripheral blood, and coinjected with human glioma cells s.c. in nude mice, promoted tumor vascularization, whereas CD14^+ Tie-2^- monocytes did not show this activity (Venneri et al., 2007).

Human TEMs are preferentially recruited to tumors, where they constitute the prominent monocyte population distinct from tumor associated macrophages (TAMs). TEMs have been observed in several mouse tumor models, including subcutaneous tumor grafts, spontaneous insulinomas developing in RIP-TAG2 transgenic mice, and human gliomas growing in the mouse brain, where they synthesize high levels of

Tumor Microenvironment Regulation of Tumor Expansion
ISBN 978-0-12-822803-6
https://doi.org/10.1016/B978-0-12-822803-6.00011-9

fibroblast growth factor-2 (FGF-2) (De Palma et al., 2003). In these tumors, TEMs constituted a small population of the total tumor infiltrating $CD11b^{+}$ myeloid cells that could be distinguished from the majority of TAMs by their surface marker profile ($Tie\text{-}2^{+}$, $Sca\text{-}1^{+}$, $CD11b^{+}$) and their proangiogenic activity (De Palma et al., 2005). TEMs were found in human tumors including those of kidney, colon, pancreas, and lung, as well as in soft tissue sarcomas, which were found both in perivascular and avascular viable (hypoxic) areas of tumors and were missing in nonneoplastic tissues adjacent to tumors (Venneri et al., 2007). Exposure to both hypoxia and angiopoietin-2 (Ang-2) markedly suppressed the release by TEMs of an antiangiogenic cytokine, namely interleukin-12 (IL-12) (De Palma et al., 2003), and of tumor necrosis factor alpha (TNF-α), which exerts a proapoptotic effect on both tumor cells and endothelial cells (Balkwill et al., 1992).

TEMs did not differentiate into endothelial cells, suggesting that their proangiogenic activity could consist of a paracrine stimulation of angiogenesis. The selective elimination of TEMs by means of a suicide gene impaired angiogenesis in mouse tumors and induced substantial tumor regression and TEM elimination did not affect the overall number of TAMs and granulocytes, indicating that TEMs represent a distinct monocyte subset with specific proangiogenic activity. Moreover, although recruited to tumors in lower numbers than TAMs, TEMs are a more potent source of proangiogenic signals. TEMs, while stimulating angiogenesis, do not actively incorporate into blood vessels and this subpopulation of $Tie\text{-}2^{+}$ cells, rather than bone marrow–derived EPCs, which are incorporated in new-forming blood vessels, promote tumor neovascularization through the release of proangiogenic factors (De Palma et al., 2003). TEMs also play a role in angiogenesis in wound healing and in nonneoplastic diseases. In mice which underwent partial hepatectomy 7–10 days earlier, TEMs were found in granulation tissue surrounding the regenerating hepatic lobules, also in proximity of newly formed vessels, suggesting that they might also contribute to promoting angiogenesis during liver regeneration (De Palma et al., 2003).

The identification of specific molecules expressed by TEMs in tumors could facilitate the design of novel anticancer therapies that selectively target these cells. De Palma and co-workers (De Palma et al., 2008), by transplanting hematopoietic progenitors transduced with a Tie-2 promoter/enhancer-driven interferon alpha (IFN-α)-1 gene, turned TEMs into IFN-α cell vehicles that targeted the IFN response to orthotopic human gliomas and spontaneous mouse mammary carcinomas and obtained significant antitumor responses and near complete abrogation of metastasis. Blockade of Ang-2 inhibited transcriptional regulation of Tie-2 in TEMs and impaired their association with tumor vasculature and their proangiogenic activity (Mazzieri et al., 2011). The selective Tie-2 inhibitor rebastinib reduced TEM infiltration in breast cancer and neuroendocrine tumors (Harney et al., 2017).

10.2 Fibroblasts

The tissue parenchyma of most organs is largely comprised of fibroblasts, that, along with the extracellular matrix (ECM) they secrete, make up the structural scaffolding of organs. Fibroblasts are primarily responsible for the synthesis, deposition, and remodeling of the ECM including collagens, fibronectin, and structural proteoglycans, as well as for the synthesis of many soluble paracrine growth factors, various classes of proteolytic enzymes and their inhibitors, and contribute to the formation of basement membrane by secreting type IV collagen and laminin (Sappino et al., 1990). Fibroblasts can be the source of endogenous inhibitors of angiogenesis, such as thrombospondin-1 (TSP-1) (Jaffe et al., 1983).

Critical elements in the tumor microenvironment include cancer associated fibroblasts (CAFs), which provide an essential communication network via secretion of growth factors and chemokines, inducing an altered ECM thereby providing additional oncogenic signals that enhance cancer cell proliferation and spread. CAFs are interconnected with tumor cells at all stages of cancer progression by promoting tumor growth, angiogenesis, and the metastatic process (Cirri & Chiarugi, 2012). In some cancers, 80 per cent of the fibroblasts within the tumor are thought to become activated (Sappino et al., 1988). CAFs differ from normal fibroblasts for increased expression of matrix proteins and abnormal secretion of proteolytic enzymes (Ohlund et al., 2016). CAFs may originate from fibroblasts which have low proliferative activity (fibrocytes) (Augsten, 2014), from epithelial cells through epithelial mesenchymal transition (EMT) (Lamouille et al., 2014), from recruited mesenchymal stem cells (MSCs) (Direkze et al., 2004), from endothelial cells through endothelial to mesenchymal transition (EndMT) (Zeisberg et al., 2007).

CAFs exhibit a higher proliferative index, as compared with normal fibroblasts, often express α-smooth muscle actin, and are surrounded by a dense accumulation of collagens (Sappino et al., 1990). Transforming growth factor beta (TGFβ) is the most important factor involved in the recruitment of CAFs and is also involved in fibroblast to myofibroblast transdifferentiation (Gabbiani, 2003; Liao & Yang, 2017). Once activated, fibroblasts differentiate, contract, proliferate, and generate excessive ECM components, leading to desmoplasia (or fibrosis) and elevated solid stress that compresses blood vessels (Stylianopoulos et al., 2018). Furthermore, desmoplasia is a physical barrier to T cell infiltration (Salmon et al., 2012) and plays a role in the resistance of metastasis to treatment (Emblem & Jain, 2016). CAFs are characterized by the expression of specific markers and secrete growth factors and angiogenic factors (Table 10.1; Table 10.2). CAFs modulate tumor growth by secreting: (i) growth factors able to increase tumor cell proliferation and exert an antiapoptotic activity and (ii) chemotactic factors able to recruit other stromal cells, including leukocytes, monocytes/macrophages, and mast cells. Moreover, CAFs (type 1 polarized fibroblasts) induce immunosuppression by an increase in Th2 cells, Th17 cells, and regulatory T cells and are also involved in therapy resistance (Müerköster et al., 2004). Coinjection of CAFs, but not of normal fibroblasts, with tumor cells resulted in enhanced tumor formation (Augsten, 2014; Hu et al., 2009). CAFs (type 2 polarized fibroblasts) are predominantly assigned with a tumor promoting function under the influence of growth factors and chemokines. They stimulate cancer cell survival, growth, and invasion, contribute to angiogenesis, and impact the activations state of various immune cells (Augsten, 2014).

Table 10.1 Markers expressed by tumor associated fibroblasts.

Vimentin
Fibroblast specific protein 1 (FSP1)
Fibroblast activating protein (FAP)
Alpha smooth actin (αSma)
Thy1 (CD90)
Thrombospondin-1 (TSP-1)
Platelet derived growth factor receptor α/β (PDGFR α/β)
Tenascin-c (TN-C)
Matrix metalloproteinase-3 (MMP-3)
Podoplanin
Neuron glial antigen 2 (NG2)
Periostin
Palladin

Table 10.2 Growth factors and angiogenic factors secreted by tumor associated fibroblasts.

Interleukin-6 (IL-6)
Insulin like growth factor (ILGF)
Hepatocyte growth factor (HGF)
Fibroblast growth factor-2 (FGF-2)
C-X-C Motif chemokine Ligand-8, -12 (CXCL-8, -12)

CAFs were increased in patients with active multiple myeloma (MM) compared to those in remission and those with monoclonal gammopathy of undetermined significance (MGUS) (Frassanito et al., 2013). The cells displayed an activated phenotype and produced high levels of TGF-β, IL-6, stromal cell-derived factor-1 alpha (SDF-1α), and insulin-like growth factor (IGF-1). Moreover, they showed a heterogeneous phenotype which entailed their origin from resident fibroblasts, and from endothelial cells and hematopoietic stem cells (HSCs) via endothelial-mesenchymal transition, and from MSCs via mesenchymal transition, all induced by both TAMs and MM plasma cells (Frassanito et al., 2013). Active MM fibroblasts favored chemotaxis, adhesion, proliferation and apoptosis resistance of MM cells through cytokines and cell-to-cell contact that was inhibited by blocking CXCR4, several integrins, and fibronectin. Studies in syngeneic 5T33MM and xenografted mouse models showed that MM cells induced the CAF recruitment and expansion which, in turn, favored tumor initiation and progression as well as angiogenesis (Frassanito et al., 2013).

CAFs produce EGF, HGF, and heregulin that promote tumor proliferation (Micke & Östman, 2005). CAF-derived CXCL-12 mediated tumor enhancement through the CXCR4 receptor expression by tumor cells and the CXCL12-CXCR4 is involved also in angiogenesis by recruiting EPCs (Orimo et al., 2005). CXCL-12 also induces VEGF expression in endothelial cells (Homey et al., 2002). CAFs promote tumor angiogenesis by secreting CXCL1 and CXCL2 to recruit proangiogenic macrophages and neutrophils into the tumor microenvironment (Erez et al., 2010). Blockade of platelet derived growth factor receptor (PDGFR) signaling in CAFs and pericytes inhibited progression of premalignant cervical lesions as well as growth of invasive carcinomas (Pietras et al., 2008). Fibroblast activating protein (FAP), selectively expressed on CAFs, exerts a proteolytic activity that supports tumor growth and proliferation (Liu et al., 2012) and has been considered as a therapeutic target in cancer. A humanized anti-FAP antibody (mAb F19; sibrotuzumab) failed to show any beneficial effect in a phase II trial for metastatic colorectal cancer (Hofheinz et al., 2003). A monoclonal anti-FAP antibody conjugated to maytansinoid induced a long-lasting inhibition of tumor growth and complete regression in different experimental tumor models, without signs of toxicity (Ostermann et al., 2008). A human therapeutic antibody 264RAD, which binds to αvβ6 and inhibits its function, delay tumor growth by preventing TGF-β-mediated activation of CAFs, and by reducing the expression of fibronectin and α-SMA on stromal fibroblasts.

10.3 Osteoblasts, osteoclasts, and adipocytes

Osteoblasts derived from MSCs following timely programmed steps requiring the expression of specific genes, including bone morphogenetic proteins (BMPs) and Wnt pathways (Grigoriadis et al., 1988). Osteoblasts express the proangiogenic factor osteopontin (OPN) and Ang-1 at the bone surface that are required for osteoblast-mediated HSCs maintenance (Arai et al., 2004). Ang-1 binds Tie2 expressed by basement membrane HSCs, inducing the activation β1 integrin and N-cadherin and contributing to the activation of HSCs (Arai et al., 2004). Moreover, Ang-1 is expressed by MM plasma cells and is associated with upregulated expression of its specific receptor Tie-2 on the bone marrow endothelial cells and enhanced angiogenesis (Giuliani et al., 2003). OPN is involved in the interplay between HSCs and osteoblast niche and inhibits HSC proliferation (Nilsson et al., 2005). Osteoblasts/HSC interactions lead to the expression of receptor activator of RANKL and Notch activation (Yin & Li, 2006). A small subset of spindle-shaped osteoblasts lining cells expressing N-cadherin is the niche for HSCs in the basement membrane, and an increase in their number is correlated with an increase in HSC number (Calvi et al., 2003) .

VEGF acts as a potent chemoattractant for osteoclasts (Mayr-wohlfart et al., 2002) that regulate VEGF expression by endothelial cells through the release of TGF-β from the bone matrix and its activation during bone resorption (Shao et al., 2009). Moreover, osteoclasts release heparinase that degrades heparin sulfate proteoglycans releasing VEGF and FGF-2, which contributes to autocrine osteoclast activation and paracrine stimulation of osteoblast activity (Collin-Osdoby et al., 2002; Padera et al., 1999; Saijo et al., 2003). Osteoclasts also release bone morphogenetic protein-7 (BMP-7), which promotes endothelial cell survival and increases VEGF production (Garimella et al., 2008), and amphiREGulin, epiREGulin, and NeuREGulin that induce endothelial cell migration and tube formation (Mehta & Besner, 2007). Secretion of matrix metalloproteinase-9 (MMP-9) by osteoclasts enhances angiogenesis through the release of VEGF from the ECM (Cackowski et al., 2010). Inhibition of osteoclasts reduces angiogenesis and tumor burden in MM (Croucher et al., 2003). Osteoclasts secrete OPN, which cooperates with VEGF of MM plasma cells to enhance angiogenesis and induce osteoclastogenic activity by endothelial cells (Tanaka et al., 2007). Moreover, in combination with IL-6, OPN enhances MM plasma cell growth (Abe et al., 2004). OPN knockout mice display minimal bone resorption compared with wild type and decreased osteoclast association at the bone surface (Asou et al., 2001). Cancer associated adipocytes produce some inflammatory cytokines, including IL-6, TNF-α, and IL-1β, which facilitate the infiltration of inflammatory cells in adipose tissue (Correa et al., 2017). Adipocytes produce a variety of angiogenic factors, including VEGF, FGF-2, and MMPs (Bouloumié et al., 2002), and obese patients exhibit elevated VEGF serum levels (Miyazawa-Hoshimoto et al., 2003).

10.4 Cancer stem cells

Cancer stem cells (CSCs) may be defined as a small subset of cancer cells within a cancer that constitute a reservoir of a self-sustaining cells with the exclusive ability to self-renew and to cause the heterogeneous lineages of cancer cells that comprise the tumor. CSCs are self-renewal cells and have the ability to generate more differentiated, nontumorigenic progeny and to initiate tumor growth upon orthotopic injection into immunocompromised mice. CSCs promote progression of cancer by escaping immune surveillance, dysregulating cellular metabolism, and inducing genomic instability. The first evidence for the existence of CSCs came from observations in acute myeloid leukemia (AML), and these cells show the same $CD34^{+}$ $CD38^{-}$ cell surface phenotype as normal HSCs. After this initial observation in AML similar hierarchical organization within tumors was shown in several other forms of leukemia, for example chronic myeloid leukemia (CML), as well as many other types of solid cancers like breast cancer, colon cancer, prostate cancer, and brain tumors (Clarke et al., 2006).

The property of self-renewal is a hallmark of CSC (Table 10.3). Mechanisms involved in self-renewal are dysregulated and lead to CSC overpopulation. Tumor cells in any cancer may be organized as a hierarchy

Table 10.3 Characteristics of cancer stem cells.

Indefinite self-renewal
Early stage differentiation
Maintains long-term growth of malignant clone
Implicated in treatment resistance and relapse

than can be roughly classified as long-term CSC (LT-CSC), short-term CSC (ST-CSC), early and late tumor progenitors, and differentiating and differentiated tumor cells. The LT-CSC should possess indefinite self-renewal and ST-CSC some self-renewal properties. CSC like normal SCs give rise to hierarchical organization of cell populations that, in turn, produce rapidly proliferating cells, resulting in the generation of fully differentiated cells (Reya et al., 2001). At the top of the hierarchy there is a CSC that expresses surface markers of primitive cells and possesses differentiation potential and limitless proliferative potential. CD133/AC133 was discovered as a marker of normal HSC and later was found to mark SC from a wide variety of tissues (Shmelkov et al., 2005).

CD133 was used to purify putative CSC in different tumors (Boman & Wicha, 2008). CD44 and CD24 have been proposed as markers for CSC several tumors; however, while some literature data indicated the existence of $CD44^{+}/CD24^{-}$ cells as CSC in breast, prostate, colon, pancreatic, and hepatic carcinoma, and melanoma (Dalerba et al., 2007; Wicha et al., 2006), other works demonstrated that the lack of CD24 expression is not an essential feature of CSC. In fact, a subpopulation of $CD24^{+}$ cells possess SCs characteristics in colon and pancreatic carcinoma (Ke et al., 2012). CD20 has been proposed in melanoma and MM; CD166 in colon carcinoma; aldehyde dehydrogenate 1A1 (ALDH1) in breast, prostate, and pancreatic carcinomas, leukemia and MM; epidermal surface antigen (ESA) in breast, colon, and pancreatic carcinomas (Boman & Wicha, 2008). ABCG2, a member of the ABC transporters, which pump a variety of endogenous and exogenous compounds out of cells, is recognized as a universal marker of SC and plays an important role in promoting SC proliferation and the maintenance of the SC phenotype. In brain tumors, CSCs express a number of neuronal SC markers, and are highly tumorigenic and resistant to radiation (Lee et al., 2006; Singh et al., 2003). CSCs also express angiogenic markers, including VEGF, VEGFR-2, Ang-1, Ang-2, and Tie2. CSC markers could be used for prediction treatment responses by means of the identification of the presence of specific CSC subtypes sensitive to the therapeutic agents and have a prognostic value by allowing assessment of the size of the CSC population within any give tumor. However, no defined localization patterns have been established when the in situ distribution of CSC markers is evaluated. In addition, the markers used to sort putative cancer-initiating cells, including CD24, CD44, and CD133, are also expressed on normal cells (Clarke et al., 2006).

The demonstration that tumor growth depends on a subpopulation of proliferating CSCs was first described in transmittable leukemias of mice. Furth and Kahn (1937) were able to transplant leukemia from one mouse to another using a single undifferentiated cell, and Makino and Kano (1955) obtained clones of tumor cells from single cells. Researchers started to search for stem-like cells in leukemias by testing the ability of various purified populations to form leukemia in nonobese diabetic/severe combined immunodeficiency (NOD-SCID) mice. Injection of leukemic cells with primitive hematopoietic progenitors resulted in leukemias that could be serially transplanted into secondary recipients, whereas injection of more differentiated leukemic cells did not. Kelly et al. (2007) used two transgenic mouse models in which the EM enhancer was used to express either the *c-myc* or *N-ras* oncogenes to induce B or T cell lymphomas. They concluded that "tumor growth need not be driven by rare cancer stem cells" based upon >10 per cent of transplanted cells give rise to tumors in syngeneic mice. The demonstration that tumor growth depends on a subpopulation of proliferating CSC was first described in transmittable leukemias of mice. Injection of leukemic cells with primitive hematopoietic progenitors resulted in leukemias that could be serially transplanted into secondary recipients, whereas injection of more differentiated leukemic cells did not. Guo et al. (2008) using a mouse model of human leukemia in which depletion of the *Pten* tumor suppressor gene in HSC resulted in a myeloproliferative disorder followed by acute T lymphoblastic leukemia demonstrated that population of leukemia SC was responsible for leukemia development. CSCs have been identified in several human solid tumors, including melanoma, brain, colorectal, hepatic, head and neck, breast, and prostate cancer (Lee et al., 2006; Reya et al., 2001; Singh et al., 2003, and are able to initiate tumor formation and metastasis, and there is a growing body of evidence suggesting that metastases develop when distant organs are seeded with CSC that arise from a primary tumor (Balic et al., 2006). Hermann et al. (2007) identified a subset of $CD133^+/CXCR4^+$ pancreatic CSCs located at the invasive front of tumors, which seem to be more responsible for metastasis. The frequency of CSCs vary among different patients and in some cases are relatively rare, whereas in others can constitute a substantial proportion of the tumor mass (Quintana et al., 2008). Glioblastoma multiforme contains neural precursors endowed with all the critical features from neural SCs (Galli et al., 2004). When neurospheres derived from highly vascularized glioblastomas were transplanted into mice, the initial tumors demonstrated low-grade glioma phenotype without any sign of angiogenesis. Upon serial transplantation, the tumor cells developed a highly malignant phenotype with extensive angiogenesis and necrosis in tumor masses (Sakariassen et al., 2006).

The abundance of CSCs differs largely between different types of tumors, different stages of the disease, as well as between individual patients. CSCs express markers of vascular endothelium and form tumor blood vessels. Brain tumor endothelial cells interact with self-renewing brain tumor cells and secrete factors that maintaining them in a "stem cell-like" state. An increased number of endothelial cells promotes the expansion of self-renewing SC and accelerates tumor growth, while their depletion has the opposite effect (Frank et al., 2010; Stupp & Hegi, 2007). Calabrese et al. (2007) established a coculture system to investigate the interaction between vascular endothelial cells and CSCs and demonstrated that endothelial cells interact with Nestin$^+$/CD133$^+$ CSC and provide factors that support their self-renewal mechanisms, demonstrating that the vasculature is essential in supporting and preserving the survival, stem-like properties, and functions of CSCs. Calabrese et al. (2007) showed that bevacizumab can reduce the stem-like cell fraction in brain tumors. Ricci-Vitani et al. (2010) demonstrated that in vitro culture of glioblastoma stem-like cells (GSCs) in conditions favorable to an endothelial cell differentiation generated a progeny with phenotypic and functional features of endothelial cells. Orthotopic or subcutaneous injection of GSC in immunocompromised mice generated large anaplastic tumor xenografts, showing a vessel wall formed by human endothelial cells derived from GSC, whereas tumor derived endothelial cells formed large anaplastic tumors in secondary recipients.

Selective targeting of endothelial cells generated by GSC in mouse xenografts resulted in tumor regression, suggesting a functional relevance of GSC-derived endothelial cells. According to these data, Wang et al. (2010) demonstrated that GSC can give rise to endothelial cells or tumor cells, through an intermediate CD133$^+$/CD144$^+$ progenitor cell. CSCs produce much higher levels of VEGF in both normal and hypoxic conditions than non-CSC population, leading to a strong angiogenic response (Bao et al., 2006). CSCs represent a plastic entity whose phenotype and function are continuously modulated by the tumor microenvironment, epigenetic regulation, and different experimental systems. Accordingly, Quintana et al. (2008) showed that modified xenotransplantation assay conditions, including the use of more highly immunocompromised NOD-SCID IL-2 receptor gamma chain null mice, can increase the detection of tumorigenic melanoma cells. CXCR4 expression by a subset of CSCs might be responsible for the signals that CSCs receive from the local microenvironment, determining which cancer cell possesses the capacity to self-renewal (Hermann et al., 2007). Moreover, CSC angiogenic potential may differ depending on the stage of tumor progression. Accordingly, more primitive cells may play a more significant role at an early stage when the tumor is smaller, whereas more mature cells may be involved as the tumor increases in size.

CSCs may reside in a vascular niche in close proximity to blood vessels (Borovski et al., 2011) named as "cancer stem cell niche" (Calabrese et al., 2007). This niche is made up of diverse stromal cells, including mesenchymal and immune cells, a vascular network, soluble factors, and components of the ECM. Relationship between CSCs and their niche may be bidirectional: CSCs generate angiogenic factors to stimulate angiogenesis, and the tumor vasculature supports maintaining CSC self-renewal and maintenance. Inhibition of vessel growth may destroy the niche and reduce the number of CSCs. Vascular niche may be considered as a specialized microenvironment that, through paracrine signaling interactions, control CSC proliferation and fate determination. The activation by angiogenic factors and inflammatory cytokines switch the vascular niche to promote tumor growth (Butler et al., 2010). Bao et al. (Bao et al., 2006) observed that GSCs $CD133^-$ are located near to blood vessels, while stem cell-like glioma cell (SCLGC)/$CD133^+$ cells generated high vascularized tumors expressing 10–20 fold more VEGF than the CD133- fraction. Moreover, anti-VEGF antibody bevacizumab depleted SCCGC-induced vascular endothelial cell migration and tube formation and inhibited the tumor growth of SCLGC-derived xenografts. Folkins et al. (2009) showed that tumor with larger SCLGC population recruited a higher amount of EPCs suggesting that SCLGCs promote angiogenesis and EPC mobilization through release of VEGF and SDF-1. Calabrese et al. (2007) demonstrated that oligodendroglioma and GSC, which expressed the major intermediate filament protein nestin, are located in close proximity to $CD34^+$ capillaries and are positively correlated to microvascular density. These findings strongly support the hypothesis that

CSC are located within a 'vascular niche.' CSCs are more resistant to chemotherapeutic agents and radiation therapy than more mature cell types from the same tissue because of increased expression of antiapoptotic proteins (Calabrese et al., 2007; Frank et al., 2010). Even if therapy kills all proliferating cells, SCs repopulate tumors. Radioresistance, chemotherapy resistance, and angiogenesis mediated by CSC could partially explain tumor recurrence in more aggressive human tumors.

Inhibition of angiogenesis alone may not be sufficient to eradicate CSCs, even if antiangiogenic therapy may sensitize CSCs to cytotoxic effects of chemotherapy. Treatment of mice transplanted with human acute myeloid leukemia (AML) cells with anti-CD44 or anti-CXCR4 antibodies markedly reduced leukemic repopulation by CSCs (Jin et al., 2006; Tavor, 2004). When these monoclonal antibodies were injected into NOD/SCID mice repopulated with AML, the leukemic SCs were markedly depleted and in some cases, the mice were cured (Jin et al., 2006; Tavor, 2004). Bone morphogenetic protein-4 (BMP-4) inhibits the tumorigenic potential of glioblastoma CSC through the BMP-4/BMPR/SMAD signaling pathway indicating that BMP plays a critical role in CSC tumorigenesis and angiogenesis (Piccirillo et al., 2006). Notch inhibitor DAPT gamma secretase inhibitor leads to a reduction of self-renewal ability of CSC and the number of $CD133^+$ tumor cells and of vascular markers, including CD105, CD31, and von Willebrand factor (Hovinga et al., 2010). It is possible to treat cancer by inducing differentiation of CSCs, i.e., differentiation therapy (Spira & Carducci, 2003). In general, SC differentiation results in two types of changes: the expression of specialized, differentiation-specific gene products, and a partial or complete restriction of the cell's capacity for further proliferation. The most thoroughly examined and clinically tested as differentiating agents are retinoic acid (RA, vitamin A), in particular all-trans-retinoic acid (ATRA). Currently, about 90 per cent of newly diagnosed patients with acute promyelocytic leukemia (APL) achieve complete remission, and over 70 per cent are cured by ATRA therapy (Camacho, 2003) with or without concomitant chemotherapy with methotrexate and cytarabine (Ohno et al., 2003). Moreover, retinoids not only induce differentiation in leukemia but also regulate antiapoptotic genes (Andreeff et al., 1999).

10.5 Circulating endothelial cells and endothelial precursor cells

Most circulating endothelial cells originate from the vessel wall and have limited growth capability, whereas a circulating endothelial progenitor subpopulation is responsible for a large majority of the endothelial cell proliferative potential. Normal adults have a low number of circulating endothelial cells per ml of peripheral blood (Solovey et al., 1997). Most of these cells are quiescent, and at least half are microvascular as defined by CD36 positivity (Solovey et al., 1997). Circulating endothelial cells are detectable in diseases marked by vascular injury, including sickle cell anemia, acute myocardial infarction, thrombotic thrombocytopenic purpura, and active cytomegalovirus infection (Grefte et al., 1993; Hladovec et al., 1978; Solovey et al., 1997).

Circulating tumor cells (CTCs) have been demonstrated as prognostic biomarkers in breast cancer, prostate cancer, and renal cell cancer (Krebs et al., 2014; Riethdorf & Pantel, 2008). Immature EPCs from primitive HSCs share common surface, i.e., AC-133, CD34, or VEGFR-2. Both EPCs and mature endothelial cells express similar endothelial-specific markers, including VEGFR-2, Tie-1, Tie-2, and VE-cadherin (Eichmann et al., 1997; Sato et al., 1993; Schnurch & Risau, 1993; Vittet et al., 1996). Moreover, HC subsets express markers similar to those of endothelial cells, including CD34, PECAM, Tie-1, Tie-2, Eph, and VEGFR-1, transcription factors such as SCL/tal-1 and AML1, vWF (Lyden et al., 2001; Suda, 2000). Asahara et al. (1997) showed that $CD34^+$, $VEGFR\text{-}2^+$ circulating EPCs form colonies that take up acetylated LDL. When $CD34^+$, $VEGFR\text{-}2^+$, $CD34^-$ or $VEGFR\text{-}2^-$ cells were injected into mice, rats, and rabbits undergoing neovascularization due to hind limb ischemia, $CD34^+$ and $VEGFR\text{-}2^+$ cells, but rarely $CD34^-$ or $VEGFR\text{-}2^-$ cells, incorporated into the vasculature in a manner consistent with their being endothelial cells (Asahara et al., 1997). AC 133 is also expressed on EPC subsets, but not on mature endothelial cells (Yin et al., 1997). Its expression is rapidly downregulated as HSCs and EPCs differentiate (Miraglia et al., 1997; Yin et al., 1997).

Peichev et al. (2000) demonstrated that a small subset of $CD34^+$ cells from different hematopoietic sources express both AC 133 and VEGFR-2. Incubation of this subset with VEGF, FGF-2, and collagen resulted in their proliferation and differentiation into AC 133^- VEGFR-2^+ mature endothelial cells. Maturation and in vitro differentiation of these cells abolish AC 133 expression, suggesting that EPCs with angioblast potential may be marked selectively with AC 133. Gehling et al. (2000) demonstrated that $AC133^+$ cells from G-CSF-mobilized peripheral blood differentiate into endothelial cells when cultured in the presence of VEGF and stem cell growth factor. Phenotypic analysis revealed that most of these cells display endothelial features, including the expression of VEGFR-2, Tie-2, and vWF. AC 133 is the best selective marker for identifying EPCs and circulating $CD34^+$ cells. VEGFR-2^+ and $AC133^+$ constitute a phenotypically and functionally distinct population of circulating endothelial cells that may play a role in postnatal vasculogenesis. AML1 is expressed in endothelial cells in sites where early HSCs emerge, such as the yolk sac (North et al., 1999), and AML1-deficient embryos, which lack definitive hematopoiesis, display defective angiogenesis in the head and pericardium (Takakura et al., 1998).

Most circulating EPCs reside in the bone marrow in close association with HSCs and the stroma. When EPCs are mobilized from adult bone marrow, they first migrate from a quiescent niche within the bone marrow into a permissive, proliferative microenvironment, the so-called vascular zone of marrow. MMP-9 as well as the survival/mitogenic activity of the stem cell cytokine SkitL are critical during this process. Release of EPCs from bone marrow into circulation can be induced by GM-CSF or VEGF and is dependent on the activity of endothelial nitric oxide synthase (eNOS) expressed by stromal cells in bone marrow (Aicher et al., 2003). Statins, erythropoietin, and estrogen also positively influence EPC number (Dimmeler et al., 2001; Strehlow et al., 2003; Vasa et al., 2001). EPCs in the peripheral blood may derive from the bone marrow and be not yet incorporated into the vessel wall. The following lines of evidence suggest that they constitute the preponderance of circulating bone marrow–derived endothelial lineage cells.

Asahara et al. (1999) employed a bone marrow transplant model in mice sublethally irradiated to demonstrate the incorporation of bone marrow EPCs into foci of neovascularization. Four weeks after transplants, when the bone marrow had been reconstituted, a variety of surgical experiments to provoke neovascularization were performed: (i) Cutaneous wounds examined 4 and 7 days after skin removal by punch biopsy disclosed a high frequency of incorporated EPCs. (ii) One week after the onset of hind limb ischemia, EPCs were incorporated in capillaries among skeletal myocytes. (iii) After permanent ligation of a coronary artery, myocardial infarction sites demonstrated incorporation of EPCs in foci of neovascularization at the border of the infarct. Shi et al. (1998) implanted an impermeable Dacron graft made in the canine thoracic aorta and demonstrated scattered islands of endothelial cells without any evidence of transmural angiogenesis. Later, they used a transplantation model in which marrow cells from donors and recipients were distinguishable to determine whether endothelial cells lining a vascular prosthesis are derived from the bone marrow (Shi et al., 1998). After 12 weeks, the graft was retrieved, and cells with endothelial morphology were identified; only donor alleles were detected in DNA from positively stained cells on the graft. These results suggested that a subset of $CD34^+$ cells located in the bone marrow mobilize to the peripheral circulation and colonize the endothelial flow surface of vascular prostheses.

Lin et al. (2000) found a distinction between vessel wall and bone marrow—derived endothelial cells in blood samples from subjects who had received gender-mismatched bone marrow transplants 5—20 months earlier. They showed that 95 percent of circulating endothelial cells had recipient genotype and 5 per cent had donor genotype. After 9 days of culture, endothelial cells derived predominantly from the recipient vessel wall, expanded only 6-fold, compared with 98-fold after 27 days by endothelial cells, mostly originated from donor bone marrow cells. These data suggest that most circulating endothelial cells in fresh blood originate from vessel walls and have limited growth capability, and that outgrowth of endothelial cells is mostly derived from transplantable marrow-derived cells. In animal models, EPCs home in on sites of active neovascularization and mobilization of bone marrow—derived EPCs, and differentiate into endothelial cells in response to tissue ischemia (Takakura et al., 1998). This finding is consistent with postnatal vasculogenesis. Identification of the chemokines that induce mobilization of bone marrow EPCs to the peripheral circulation may provide a novel mechanism for their recruitment to sites of vascular trauma to accelerate vascular healing (Rafii, 2000). Moreover, expansion and mobilization of EPCs may augment the resident population of endothelial cells competent to respond to exogenous angiogenic cytokines (Isner & Asahara, 1999). Kalka et al. (2000) demonstrated that transplantation of EPCs to athymic mice with hind limb ischemia markedly improves blood flow recovery and capillary density in the ischemic limb and significantly reduces the rate of limb loss. The recruitment of EPCs to tumor angiogenesis is a multistep process, including: (i) active arrest and homing of the circulating cells within the angiogenic microvasculature; (ii) transendothelial extravasation into the interstitial space; (iii) extravascular formation of cellular clusters; (iv) creation of vascular sprouts and cellular networks; (v) incorporation into a functional microvasculature. High levels of VEGF produced by tumors may result in the mobilization of bone marrow–derived SCs in the peripheral circulation and enhance their recruitment into the tumor vasculature (Asahara et al., 1999; Hattori et al., 2001). Moreover, Hattori et al. (2001) showed that combined elevation of VEGF and Ang-1 result in remodeling of the bone architecture, with depletion of the sinusoidal spaces of hematopoietic cells and a parallel increase in bone marrow vascularization and splenomegaly. Lyden et al. (2001) demonstrated that transplantation and engraftment of b-galactosidase-positive wild-type bone marrow or VEGF-mobilized stem cells into lethally irradiated Id-mutant mice is sufficient to reconstitute tumor angiogenesis. In contrast to wild-type mice, Id-mutant mice fail to support the growth of tumors because of impaired angiogenesis. Tumor analysis demonstrates the uptake of bone marrow–derived VEGFR-2$^+$ EPCs into vessels surrounded by VEGFR-1$^+$ myeloid cells. Defective angiogenesis in Id-mutant mice is associated with impaired VEGF-induced mobilization and proliferation of the bone marrow precursor cells. Inhibition of both VEGFR-1 and VEGFR-2 signaling is needed to block tumor angiogenesis and induce necrosis. Vajkoczy et al. (2003) proposed the term "angiomorphosis" as the process of enhancing tissue vascularization through active recruitment of EPCs from the circulation by endothelial cells and the action of EPCs as organizers of the angiogenic process. EPC levels correlate with poor overall survival in non small cell lung cancer (NSCLC) and patients with partial or complete remission after therapy present lower levels of circulating EPCs than patients with stable or progressive disease, suggesting that EPC levels correlate with the efficacy of treatment (Dome et al., 2006).

10.5.1 Multipotent adult progenitor cells (MAPCs) and myeloid-derived suppressor cells (MDSCs)

Reyes et al. (2002) have identified a single cell in human and rodent postnatal marrow that they termed MAPCs. MAPCs were selected by depleting adult bone marrow of HC expressing CD45 and glycophorin-A, followed by long-term culture on fibronectin with epidermal growth factor (EGF) and platelet derived growth factor (PDGF) under low serum conditions. A cell population expressing AC133 and low levels of VEGFR-1 as well as VEGFR-2 and the embryonic stem cell marker Oct-4 emerged.

Its culture with VEGF induced differentiation into $CD34^+$, VE-cadherin$^+$, VEGFR-2$^+$ cells, a phenotype consistent with angioblasts. Subsequently these cells express vWF and markers of mature endothelium, such as CD31, CD36, and CD62-P. MAPCs form vascular tubes when plated on Matrigel and upregulate angiogenic receptors and VEGF in response to hypoxia. MDSCs have been characterized by cell surface coexpression of CD11b and Gr1 (Gabrilovich & Nagaraj, 2009).

MDSCs are precursors of dendritic cells, macrophages, and granulocytes, and they can differentiate into protumoral phenotype in hypoxia condition (Sica & Massarotti, 2017). In cancer, MDSCs undergo abnormal expansion and accumulate in different sites, including bone marrow, spleen, liver, and tumor site sustaining tumor-derived inflammation and inducing tumor progression, neovascularization, and immune escape. MDSCs are mobilized during tumorigenesis and infiltrate developing tumors where they promote vascularization and disrupt major mechanisms of immunosurveillance (Gabrilovich et al., 2012). An elevated number of peripheral MDSCs correlate with advance disease and therapeutic inefficacy (Diaz-Montero et al., 2009). MDSCs are one of the major components of the immunosuppressive network responsible for the impairment of the natural killer (NK) cell and T cell—dependent anticancer immunity and promote angiogenesis, cancer cell invasion, and metastasis (Gabrilovich & Nagaraj, 2009; Monu & Frey, 2012). The immunosuppressive effects of MDSCs are exerted through activation and expansion of the Treg population, interference with lymphocyte trafficking, creation of oxidative stress, and depletion of nutrients (Gabrilovich et al., 2012).

GM-CSF, VEGF, and IL-3 promote blockade of myeloid maturation with an expansion in the MDSC population. Combined treatment with anti-Gr1 and anti-VEGF results in more pronounced reduction in the vascular surface area compared with anti-VEGF treatment alone (Shojaei et al., 2007). MDSCs and Treg cells are mobilized into the circulation in response to activated cytokines that are induced by tumorigenesis, including TGF-β and CXCL5-CXCR2. MDSCs and Treg cells infiltrate the growing tumor to disrupt immune surveillance through multiple mechanisms, including disruption of antigen factors, secrete ECM proteins and basement membrane components, regulate differentiation, modulate immune responses, and contribute to deregulated homeostasis. The identification of molecular pathways involved in MDSC function will lead to the development of new agents that disrupt the tumor—host interactions, thus improving the efficacy of immunotherapy.

References

Abe, M., Hiura, K., Wilde, J., Shioyasono, A., Moriyama, K., Hashimoto, T., et al. (2004). Osteoclasts enhance myeloma cell growth and survival via cell-cell contact: A vicious cycle between bone destruction and myeloma expansion. *Blood, 104*(8), 2484–2491. https://doi.org/10.1182/blood-2003-11-3839.

Aicher, A., Heeschen, C., Mildner-Rihm, C., Urbich, C., Ihling, C., Technau-Ihling, K., et al. (2003). Essential role of endothelial nitric oxide synthase for mobilization of stem and progenitor cells. *Nature Medicine, 9*(11), 1370–1376. https://doi.org/10.1038/nm948.

Andreeff, M., Jiang, S., Zhang, X., Konopleva, M., Estrov, Z., Snell, V. E., et al. (1999). Expression of Bcl-2-related genes in normal and AML progenitors: Changes induced by chemotherapy and retinoic acid. *Leukemia, 13*(11), 1881–1892. https://doi.org/10.1038/sj.leu.2401573.

Arai, F., Hirao, A., Ohmura, M., Sato, H., Matsuoka, S., Takubo, K., et al. (2004). Tie2/Angiopoietin-1 signaling regulates hematopoietic stem cell quiescence in the bone marrow niche. *Cell, 118*(2), 149–161. https://doi.org/10.1016/j.cell.2004.07.004.

Asahara, T., Murohara, T., Sullivan, A., Silver, M., van der Zee, R., Li, T., et al. (1997). Isolation of putative progenitor endothelial cells for angiogenesis. *Science, 275*(5302), 964–967. https://doi.org/10.1126/science.275.5302.964.

Asahara, T., Takahashi, T., Masuda, H., Kalka, C., Chen, D., Iwaguro, H., et al. (1999). VEGF contributes to postnatal neovascularization by mobilizing bone marrow-derived endothelial progenitor cells. *EMBO Journal, 18*(14), 3964–3972. https://doi.org/10.1093/emboj/18.14.3964.

Asou, Y., Rittling, S. R., Yoshitake, H., Tsuji, K., Shinomiya, K., Nifuji, A., et al. (2001). Osteopontin facilitates angiogenesis, accumulation of osteoclasts, and resorption in ectopic bone. *Endocrinology, 142*(3), 1325–1332. https://doi.org/10.1210/endo.142.3.8006.

Augsten, M. (2014). Cancer-associated fibroblasts as another polarized cell type of the tumor microenvironment. *Frontiers in Oncology, 4:62.* https://doi.org/10.3389/fonc.2014.00062. eCollection 2014.

Balic, M., Lin, H., Young, L., Hawes, D., Giuliano, A., McNamara, G., et al. (2006). Most early disseminated cancer cells detected in bone marrow of breast cancer patients have a putative breast cancer stem cell phenotype. *Clinical Cancer Research, 12*(19), 5615–5621. https://doi.org/10.1158/1078-0432.CCR-06-0169.

Balkwill, F. R., Bonnet, D., & Dick, J. (1992). Human acute myeloid leukemia is organized as a hierarchy that originates from a primitive hematopoietic cell. *Progress in Growth Factors Research, 4*, 730–737.

Bao, S., Wu, Q., Sathornsumetee, S., Hao, Y., Li, Z., Hjelmeland, A. B., et al. (2006). Stem cell–like glioma cells promote tumor angiogenesis through vascular endothelial growth factor. *Cancer Research, 66*(16), 7843–7848. https://doi.org/10.1158/0008-5472.can-06-1010.

Boman, B. M., & Wicha, M. S. (2008). Cancer stem cells: A step toward the cure. *Journal of Clinical Oncology, 26*(17), 2795–2799. https://doi.org/10.1200/JCO.2008.17.7436.

Borovski, T., De Sousa E Melo, F., Vermeulen, L., & Medema, J. P. (2011). Cancer stem cell niche: The place to be. *Cancer Research, 71*(3), 634–639. https://doi.org/10.1158/0008-5472.CAN-10-3220.

Bouloumié, A., Lolmède, K., Sengenès, C., Galitzky, J., & Lafontan, M. (2002). Angiogenesis in adipose tissue. In *Annales d'Endocrinologie, 63* pp. 91–95). Elsevier Masson SAS. http://www.elsevier.com/wps/find/journaldescription.cws_home/709545/description#description. Issue 2 I.

Butler, J. M., Kobayashi, H., & Rafii, S. (2010). Instructive role of the vascular niche in promoting tumour growth and tissue repair by angiocrine factors. *Nature Reviews Cancer, 10*(2), 138–146. https://doi.org/10.1038/nrc2791.

Cackowski, F. C., Anderson, J. L., Patrene, K. D., Choksi, R. J., Shapiro, S. D., Windle, J. J., et al. (2010). Osteoclasts are important for bone angiogenesis. *Blood, 115*(1), 140–149. https://doi.org/10.1182/blood-2009-08-237628.

Calabrese, C., Poppleton, H., Kocak, M., Hogg, T. L., Fuller, C., Hamner, B., et al. (2007). A perivascular niche for brain tumor stem cells. *Cancer Cell, 11*(1), 69–82. https://doi.org/10.1016/j.ccr.2006.11.020.

Calvi, L. M., Adams, G. B., Weibrecht, K. W., Weber, J. M., Olson, D. P., Knight, M. C., et al. (2003). Osteoblastic cells regulate the haematopoietic stem cell niche. *Nature, 425*(6960), 841–846. https://doi.org/10.1038/nature02040.

Camacho, L. H. (2003). Clinical applications of retinoids in cancer medicine. *Journal of Biological Regulators and Homeostatic Agents, 17*(1), 98–114. http://www.biolifesas.org/contentsJBRHA.htm.

Cirri, P., & Chiarugi, P. (2012). Cancer-associated-fibroblasts and tumour cells: A diabolic liaison driving cancer progression. *Cancer and Metastasis Reviews, 31*(1–2), 195–208. https://doi.org/10.1007/s10555-011-9340-x.

Clarke, M. F., Dick, J. E., Dirks, P. B., Eaves, C. J., Jamieson, C. H. M., Jones, D. L., et al. (2006). Cancer stem cells - Perspectives on current status and future directions: AACR workshop on cancer stem cells. *Cancer Research, 66*(19), 9339–9344. https://doi.org/10.1158/0008-5472.CAN-06-3126.

Collin-Osdoby, P., Rothe, L., Bekker, S., Anderson, F., Huang, Y., & Osdoby, P. (2002). Basic fibroblast growth factor stimulates osteoclast recruitment, development, and bone pit resorption in association with angiogenesis in vivo on the chick chorioallantoic membrane and activates isolated avian osteoclast resorption in vitro. *Journal of Bone and Mineral Research, 17*(10), 1859–1871. https://doi.org/10.1359/jbmr.2002.17.10.1859.

Correa, L., Correa, R., & Farinasso, C. (2017). Adipocytes and macrophages interplay in the orchestration of tumor microenvironment: New implications in cancer progression. *Frontiers in Immunology, 19*(8): 1129. https://doi.org/10.3389/fimmu.2017.01129. eCollection 2017.

Croucher, P. I., De Raeve, H., Perry, M. J., Hijzen, A., Shipman, C. M., Lippitt, J., et al. (2003). Zoledronic acid treatment of 5T2MM-bearing mice inhibits the development of myeloma bone disease: Evidence for decreased osteolysis, tumor burden and angiogenesis, and increased survival. *Journal of Bone and Mineral Research, 18*(3), 482–492. https://doi.org/10.1359/jbmr.2003.18.3.482.

Dalerba, P., Cho, R. W., & Clarke, M. F. (2007). Cancer stem cells: Models and concepts. *Annual Review of Medicine, 58*, 267–284. https://doi.org/10.1146/annurev.med.58.062105.204854.

De Palma, M., Venneri, M. A., Galli, R., SSergi, L., Politi, L. S., Sampaolesi, M., & Naldini, L. (2005). Tie2 identifies a hematopoietic lineage of proangiogenic monocytes required for tumor vessel formation and a mesenchymal population of pericyte progenitor. *Cancer Cell, 8*, 211–226.

De Palma, M., Venneri, M. A., Roca, C., & Naldini, L. (2003). Targeting exogenous genes to tumor angiogenesis by transplantation of genetically modified hematopoietic stem cells. *Nature Medicine, 9*(6), 789–795. https://doi.org/10.1038/nm871.

De Palma, M., Mazzieri, R., Politi, L. S., Pucci, F., Zonari, E., Sitia, G., et al. (2008). Tumor-targeted interferon-α delivery by Tie2-expressing monocytes inhibits tumor growth and metastasis. *Cancer Cell, 14*(4), 299–311. https://doi.org/10.1016/j.ccr.2008.09.004.

Diaz-Montero, C. M., Salem, M. L., Nishimura, M. I., Garrett-Mayer, E., Cole, D. J., & Montero, A. J. (2009). Increased circulating myeloid-derived suppressor cells correlate with clinical cancer stage, metastatic tumor burden, and doxorubicin-cyclophosphamide chemotherapy. *Cancer Immunology, Immunotherapy, 58*(1), 49–59. https://doi.org/10.1007/s00262-008-0523-4.

Dimmeler, S., Aicher, A., Vasa, M., Mildner-Rihm, C., Adler, K., Tiemann, M., et al. (2001). HMG-CoA reductase inhibitors (statins) increase endothelial progenitor cells via the PI 3-kinase/Akt pathway. *Journal of Clinical Investigation, 108*(3), 391–397. https://doi.org/10.1172/jci200113152.

Direkze, N. C., Hodivala-Dilke, K., Jeffery, R., Hunt, T., Poulsom, R., Oukrif, D., et al. (2004). Bone marrow contribution to tumor-associated myofibroblasts and fibroblasts. *Cancer Research, 64*(23), 8492–8495. https://doi.org/10.1158/0008-5472.can-04-1708.

Dome, B., Timar, J., Dobos, J., Meszaros, L., Raso, E., Paku, S., et al. (2006). Identification and clinical significance of circulating endothelial progenitor cells in human non–small cell lung cancer. *Cancer Research, 66*(14), 7341–7347. https://doi.org/10.1158/0008-5472.can-05-4654.

Eberlein, C., Kendrew, J., McDaid, K., Alfred, A., Kang, J. S., Jacobs, V. N., et al. (2012). A human monoclonal antibody 264RAD targeting αvβ6 integrin reduces tumour growth and metastasis, and modulates key biomarkers in vivo. *Oncogene, 32*(37), 4406–4416. https://doi.org/10.1038/onc.2012.460.

Eichmann, A., Corbel, C., Nataf, V., Vaigot, P., Bréant, C., & Le Douarin, N. M. (1997). Ligand-dependent development of the endothelial and hemopoietic lineages from embryonic mesodermal cells expressing vascular endothelial growth factor receptor 2. *Proceedings of the National Academy of Sciences of the United States of America, 94*(10), 5141–5146. https://doi.org/10.1073/pnas.94.10.5141.

Emblem, K. E., & Jain, R. K. (2016). Improving treatment of liver metastases by targeting nonangiogenic mechanisms. *Nature Medicine, 22*(11), 1209–1210. https://doi.org/10.1038/nm.4228.

Erez, N., Truitt, M., Olson, P., & Hanahan, D. (2010). Cancer-associated fibroblasts are activated in incipient neoplasia to orchestrate tumor-promoting inflammation in an NF-κB-dependent manner. *Cancer Cell, 17*(2), 135–147. https://doi.org/10.1016/j.ccr.2009.12.041.

Folkins, C., Shaked, Y., Man, S., Tang, T., Lee, C. R., Zhu, Z., et al. (2009). Glioma tumor stem-like cells promote tumor angiogenesis and vasculogenesis via vascular endothelial growth factor and stromal-derived factor 1. *Cancer Research, 69*(18), 7243–7251. https://doi.org/10.1158/0008-5472.CAN-09-0167.

Frank, N. Y., Schatton, T., & Frank, M. H. (2010). The therapeutic promise of the cancer stem cell concept. *Journal of Clinical Investigation, 120*(1), 41–50. https://doi.org/10.1172/JCI41004.

Frassanito, M. A., Rao, L., Moschetta, M., Ria, R., Di Marzo, L., De Luisi, A., et al. (2013). Bone marrow fibroblasts parallel multiple myeloma progression in patients and mice: In vitro and in vivo studies. *Leukemia, 28*(4), 904–916. https://doi.org/10.1038/leu.2013.254.

Furth, J., & Kahn, M. C. (1937). The transmission of leukemia of mice with a single cell. *American Journal of Cancer, 31*(2), 276–282. https://doi.org/10.1158/ajc.1937.276.

Gabbiani, G. (2003). The myofibroblast in wound healing and fibrocontractive diseases. *Journal of Pathology, 200*(4), 500–503. https://doi.org/10.1002/path.1427.

Gabrilovich, D. I., & Nagaraj, S. (2009). Myeloid-derived suppressor cells as regulators of the immune system. *Nature Reviews Immunology, 9*(3), 162–174. https://doi.org/10.1038/nri2506.

Gabrilovich, D. I., Ostrand-Rosenberg, S., & Bronte, V. (2012). Coordinated regulation of myeloid cells by tumours. *Nature Reviews Immunology, 12*(4), 253–268. https://doi.org/10.1038/nri3175.

Galli, R., Binda, E., Orfanelli, U., Cipelletti, B., Gritti, A., De Vitis, S., et al. (2004). Isolation and characterization of tumorigenic, stem-like neural precursors from human glioblastoma. *Cancer Research, 64*(19), 7011–7021. https://doi.org/10.1158/0008-5472.CAN-04-1364.

Garimella, R., Tague, S. E., Zhang, J., Belibi, F., Nahar, N., Sun, B. H., et al. (2008). Expression and synthesis of bone morphogenetic proteins by osteoclasts: A possible path to anabolic bone remodeling. *Journal of Histochemistry and Cytochemistry, 56*(6), 569–577. https://doi.org/10.1369/jhc.2008.950394.

Gehling, U. M., Ergün, S., Schumacher, U., Wagener, C., Pantel, K., Otte, M., et al. (2000). In vitro differentiation of endothelial cells from AC133-positive progenitor cells. *Blood, 95*(10), 3106–3112. https://doi.org/10.1182/blood.v95.10.3106.010k08_3106_3112.

Giuliani, N., Colla, S., Lazzaretti, M., Sala, R., Roti, G., Mancini, C., et al. (2003). Proangiogenic properties of human myeloma cells: Production of angiopoietin-1 and its potential relationship to myeloma-induced angiogenesis. *Blood, 102*(2), 638–645. https://doi.org/10.1182/blood-2002-10-3257.

Grefte, A., Giessen, M. V.d., Son, W. V., & The, T. H. (1993). Circulating cytomegalovirus (CMV)-infected endothelial cells in patients with an active CMV infection. *Journal of Infectious Diseases, 167*(2), 270–277. https://doi.org/10.1093/infdis/167.2.270.

Grigoriadis, A. E., Heersche, J. N. M., & Aubin, J. E. (1988). Differentiation of muscle, fat, cartilage, and bone from progenitor cells present in a bone-derived clonal cell population: Effect of dexamethasone. *Journal of Cell Biology, 106*(6), 2139–2151. https://doi.org/10.1083/jcb.106.6.2139.

Guo, W., Lasky, J. L., Chang, C.-J., Mosessian, S., Lewis, X., Xiao, Y., et al. (2008). Multi-genetic events collaboratively contribute to Pten-null leukaemia stem-cell formation. *Nature, 453*(7194), 529–533. https://doi.org/10.1038/nature06933.

Harney, A. S., Karagiannis, G. S., Pignatelli, J., Smith, B. D., Kadioglu, E., Wise, S. C., et al. (2017). The selective Tie2 inhibitor rebastinib blocks recruitment and function of Tie2(Hi) macrophages in breast cancer and pancreatic neuroendocrine tumors. *Molecular Cancer Therapeutics, 16*(11), 2486–2501. https://doi.org/10.1158/1535-7163.MCT-17-0241.

Hattori, K., Dias, S., Heissig, B., Hackett, N. R., Lyden, D., Tateno, M., et al. (2001). Vascular endothelial growth factor and angiopoietin-1 stimulate postnatal hematopoiesis by recruitment of vasculogenic and hematopoietic stem cells. *The Journal of Experimental Medicine, 193*(9), 1005–1014. https://doi.org/10.1084/jem.193.9.1005.

Hermann, P. C., Huber, S. L., Herrler, T., Aicher, A., Ellwart, J. W., Guba, M., et al. (2007). Distinct populations of cancer stem cells determine tumor growth and metastatic activity in human pancreatic cancer. *Cell Stem Cell, 1*(3), 313–323. https://doi.org/10.1016/j.stem.2007.06.002.

Hladovec, J., Přerovský, I., Staněk, V., & Fabián, J. (1978). Circulating endothelial cells in acute myocardial infarction and angina pectoris. *Klinische Wochenschrift, 56*(20), 1033–1036. https://doi.org/10.1007/bf01476669.

Hofheinz, R. D., Weisser, A., Willer, A., Hehlmann, R., & Hochhaus, A. (2003). Treatment of a patient with advanced esophageal cancer with a combination of mitomycin C and capecitabine: Activation of the thymidine phosphorylase as active principle? *Onkologie, 26*(2), 161–164. https://doi.org/10.1159/000069837.

Homey, B., Müller, A., & Zlotnik, A. (2002). Chemokines: Agents for the immunotherapy of cancer? *Nature Reviews Immunology, 2*(3), 175–184. https://doi.org/10.1038/nri748.

Hovinga, K. E., Shimizu, F., Wang, R., Panagiotakos, G., Van Der Heijden, M., Moayedpardazi, H., et al. (2010). Inhibition of notch signaling in glioblastoma targets cancer stem cells via an endothelial cell intermediate. *Stem Cells (Dayton, Ohio), 28*(6), 1019–1029. https://doi.org/10.1002/stem.429.

Hu, M., Peluffo, G., Chen, H., Gelman, R., Schnitt, S., & Polyak, K. (2009). Role of COX-2 in epithelial-stromal cell interactions and progression of ductal carcinoma in situ of the breast. *Proceedings of the National Academy of Sciences of the United States of America, 106*(9), 3372–3377. https://doi.org/10.1073/pnas.0813306106.

Isner, J. M., & Asahara, T. (1999). Angiogenesis and vasculogenesis as therapeutic strategies for postnatal neovascularization. *Journal of Clinical Investigation, 103*(9), 1231–1236. https://doi.org/10.1172/JCI6889.

Jaffe, E. A., Ruggiero, J. T., Leung, L. L. K., Doyle, M. J., McKeown-Longo, P. J., & Mosher, D. F. (1983). Cultured human fibroblasts synthesize and secrete thrombospondin and incorporate it into extracellular matrix. *Proceedings of the National Academy of Sciences of the United States of America, 80*(4 I), 998–1002. https://doi.org/10.1073/pnas.80.4.998.

Jin, L., Hope, K. J., Zhai, Q., Smadja-Joffe, F., & Dick, J. E. (2006). Targeting of CD44 eradicates human acute myeloid leukemic stem cells. *Nature Medicine, 12*(10), 1167–1174. https://doi.org/10.1038/nm1483.

Jones, N., Iljin, K., Dumont, D. J., & Alitalo, K. (2001). Tie receptors: New modulators of angiogenic and lymphangiogenic responses. *Nature Reviews Molecular Cell Biology, 2*(4), 257–267. https://doi.org/10.1038/35067005.

Kalka, C., Masuda, H., Takahashi, T., Gordon, R., Tepper, O., Gravereaux, E., et al. (2000). Vascular endothelial growth factor165 gene transfer augments circulating endothelial progenitor cells in human subjects. *Circulation Research, 86*(12), 1198–1202. https://doi.org/10.1161/01.res.86.12.1198.

Ke, J., Wu, X., Wu, X., He, X., Lian, L., Zou, Y., et al. (2012). A subpopulation of CD24+ cells in colon cancer cell lines possess stem cell characteristics. *Neoplasma, 59*(03), 282–288. https://doi.org/10.4149/neo_2012_036.

Kelly, P. N., Dakic, A., Adams, J. M., Nutt, S. L., & Strasser, A. (2007). Tumor growth need not be driven by rare cancer stem cells. *Science, 317*(5836), 337. https://doi.org/10.1126/science.1142596.

Krebs, M. G., Metcalf, R. L., Carter, L., Brady, G., Blackhall, F. H., & Dive, C. (2014). Molecular analysis of circulating tumour cells - Biology and biomarkers. *Nature Reviews Clinical Oncology, 11*(3), 129–144. https://doi.org/10.1038/nrclinonc.2013.253.

Lamouille, S., Xu, J., & Derynck, R. (2014). Molecular mechanisms of epithelial-mesenchymal transition. *Nature Reviews Molecular Cell Biology, 15*(3), 178–196. https://doi.org/10.1038/nrm3758.

Lee, J., Kotliarova, S., Kotliarov, Y., Li, A., Su, Q., Donin, N. M., et al. (2006). Tumor stem cells derived from glioblastomas cultured in bFGF and EGF more closely mirror the phenotype and genotype of primary tumors than do serum-cultured cell lines. *Cancer Cell, 9*(5), 391–403. https://doi.org/10.1016/j.ccr.2006.03.030.

Liao, T. T., & Yang, M. H. (2017). Revisiting epithelial-mesenchymal transition in cancer metastasis: The connection between epithelial plasticity and stemness. *Molecular Oncology, 11*(7), 792–804. https://doi.org/10.1002/1878-0261.12096.

Lin, Y., Weisdorf, D. J., Solovey, A., & Hebbel, R. P. (2000). Origins of circulating endothelial cells and endothelial outgrowth from blood. *Journal of Clinical Investigation, 105*(1), 71–77. https://doi.org/10.1172/JCI8071.

Liu, R., Li, H., Liu, L., Yu, J., & Ren, X. (2012). Fibroblast activation protein: A potential therapeutic target in cancer. *Cancer Biology and Therapy, 13*(3), 123–129. https://doi.org/10.4161/cbt.13.3.18696.

Lyden, D., Hattori, K., Dias, S., Costa, C., Blaikie, P., Butros, L., et al. (2001). Impaired recruitment of bone-marrow–derived endothelial and hematopoietic precursor cells blocks tumor angiogenesis and growth. *Nature Medicine,* 7(11), 1194–1201. https://doi.org/10.1038/nm1101-1194.

Makino, S., & Kano, K. (1955). Cytological studies of tumors. xiv. isolation of single-cell clones from a mixed-cell tumor of the rat. *Journal of the National Cancer Institute, 15*(5), 1165–1181. https://doi.org/10.1093/jnci/15.5.1165.

Mayr-wohlfart, U., Waltenberger, J., Hausser, H., Kessler, S., Günther, K. P., Dehio, C., et al. (2002). Vascular endothelial growth factor stimulates chemotactic migration of primary human osteoblasts. *Bone, 30*(3), 472–477. https://doi.org/10.1016/s8756-3282(01)00690-1.

Mazzieri, R., Pucci, F., Moi, D., Zonari, E., Ranghetti, A., Berti, A., ... De Palma, M. (2011). Targeting the ANG2/TIE2 axis inhibits tumor growth and metastasis by impairing angiogenesis and disabling rebounds of proangiogenic myeloid cells. *Cancer Cell, 19*(4), 512–526.

Mehta, V. B., & Besner, G. E. (2007). HB-EGF promotes angiogenesis in endothelial cells via PI3-kinase and MAPK signaling pathways. *Growth Factors, 25*(4), 253–263. https://doi.org/10.1080/08977190701773070.

Micke, P., & Östman, A. (2005). Exploring the tumour environment: Cancer-associated fibroblasts as targets in cancer therapy. *Expert Opinion on Therapeutic Targets, 9*(6), 1217–1233. https://doi.org/10.1517/14728222.9.6.1217.

Miraglia, S., Godfrey, W., Yin, A. H., Atkins, K., Warnke, R., Holden, J. T., et al. (1997). A novel five-transmembrane hematopoietic stem cell antigen: Isolation, characterization, and molecular cloning. *Blood, 90*(12), 5013–5021. https://doi.org/10.1182/blood.v90.12.5013.5013_5013_5021

Miyazawa-Hoshimoto, S., Takahashi, K., Bujo, H., Hashimoto, N., & Saito, Y. (2003). Elevated serum vascular endothelial growth factor is associated with visceral fat accumulation in human obese subjects. *Diabetologia, 46*(11), 1483–1488. https://doi.org/10.1007/s00125-003-1221-6.

Monu, N. R., & Frey, A. B. (2012). Myeloid-derived suppressor cells and anti-tumor T cells: A complex relationship. *Immunological Investigations, 41*(6–7), 595–613. https://doi.org/10.3109/08820139.2012.673191.

Müerköster, S., Wegehenkel, K., Arlt, A., Witt, M., Sipos, B., Kruse, M.-L., et al. (2004). Tumor stroma interactions induce chemoresistance in pancreatic ductal carcinoma cells involving increased secretion and paracrine effects of nitric oxide and interleukin-1β. *Cancer Research, 64*(4), 1331–1337. https://doi.org/10.1158/0008-5472.can-03-1860.

Nilsson, S. K., Johnston, H. M., Whitty, G. A., Williams, B., Webb, R. J., Denhardt, D. T., et al. (2005). Osteopontin, a key component of the hematopoietic stem cell niche and regulator of primitive hematopoietic progenitor cells. *Blood, 106*(4), 1232–1239. https://doi.org/10.1182/blood-2004-11-4422.

North, T., Gu, T. L., Stacy, T., Wang, Q., Howard, L., Binder, M., et al. (1999). Cbfa2 is required for the formation of intra-aortic hematopoietic clusters. *Development, 126*(11), 2563–2575.

Öhlund, D., Handly-Santana, A., Elyada, E., Biffi, G., & Tuveson, D. A. (2016). Abstract B73: Pancreatic stellate cell heterogeneity in cancer. In *Heterogeneity of Pancreatic Cancer*. American Association for Cancer Research. https://doi.org/10.1158/1538-7445.panca16-b73.

Ohno, R., Asou, N., & Ohnishi, K. (2003). Treatment of acute promyelocytic leukemia: Strategy toward further increase of cure rate. *Leukemia, 17*(8), 1454–1463. https://doi.org/10.1038/sj.leu.2403031.

Orimo, A., Gupta, P. B., Sgroi, D. C., Arenzana-Seisdedos, F., Delaunay, T., Naeem, R., et al. (2005). Stromal fibroblasts present in invasive human breast carcinomas promote tumor growth and angiogenesis through elevated SDF-1/CXCL12 secretion. *Cell, 121*(3), 335–348. https://doi.org/10.1016/j.cell.2005.02.034.

Ostermann, E., Garin-Chesa, P., Heider, K. H., Kalat, M., Lamche, H., Puri, C., et al. (2008). Effective immunoconjugate therapy in cancer models targeting a serine protease of tumor fibroblasts. *Clinical Cancer Research, 14*(14), 4584–4592. https://doi.org/10.1158/1078-0432.CCR-07-5211.

Padera, R., Venkataraman, G., Berry, D., Godavarti, R., & Sasisekharan, R. (1999). FGF-2/fibroblast growth factor receptor/heparin-like glycosaminoglycan interactions: A compensation model for FGF-2 signaling. *FASEB Journal, 13*(13), 1677–1687. https://doi.org/10.1096/fasebj.13.13.1677.

Peichev, M., Naiyer, A. J., Pereira, D., Zhu, Z., Lane, W. J., Williams, M., et al. (2000). Expression of VEGFR-2 and AC133 by circulating human CD34+ cells identifies a population of functional endothelial precursors. *Blood, 95*(3), 952–958. https://doi.org/10.1182/blood.v95.3.952.003k27_952_958.

Piccirillo, S. G. M., Reynolds, B. A., Zanetti, N., Lamorte, G., Binda, E., Broggi, G., et al. (2006). Bone morphogenetic proteins inhibit the tumorigenic potential of human brain tumour-initiating cells. *Nature, 444*(7120), 761–765. https://doi.org/10.1038/nature05349.

Pietras, K., Pahler, J., Bergers, G., & Hanahan, D. (2008). Functions of paracrine PDGF signaling in the proangiogenic tumor stroma revealed by pharmacological targeting. *PLoS Medicine, 5*(1), 0123–0138. https://doi.org/10.1371/journal.pmed.0050019.

Quintana, E., Shackleton, M., Sabel, M. S., Fullen, D. R., Johnson, T. M., & Morrison, S. J. (2008). Efficient tumour formation by single human melanoma cells. *Nature, 456*(7222), 593–598. https://doi.org/10.1038/nature07567.

Rafii, S. (2000). Circulating endothelial precursors: Mystery, reality, and promise. *Journal of Clinical Investigation, 105*(1), 17–19. https://doi.org/10.1172/JCI8774.

Reya, T., Morrison, S. J., Clarke, M. F., & Weissman, I. L. (2001). Stem cells, cancer, and cancer stem cells. *Nature, 414*(6859), 105–111. https://doi.org/10.1038/35102167.

Reyes, M., Dudek, A., Jahagirdar, B., Koodie, L., Marker, P. H., & Verfaillie, C. M. (2002). Origin of endothelial progenitors in human postnatal bone marrow. *Journal of Clinical Investigation, 109*(3), 337–346. https://doi.org/10.1172/JCI0214327.

Ricci-Vitiani, L., Pallini, R., Biffoni, M., Todaro, M., Invernici, G., Cenci, T., et al. (2010). Tumour vascularization via endothelial differentiation of glioblastoma stem-like cells. *Nature, 468*(7325), 824–828. https://doi.org/10.1038/nature09557.

Riethdorf, S., & Pantel, K. (2008). Disseminated tumor cells in bone marrow and circulating tumor cells in blood of breast cancer patients: Current state of detection and characterization. *Pathobiology, 75*(2), 140–148. https://doi.org/10.1159/000123852.

Saijo, M., Kitazawa, R., Nakajima, M., Kurosaka, M., Maeda, S., & Kitazawa, S. (2003). Heparanase mRNA expression during fracture repair in mice. *Histochemistry and Cell Biology, 120*(6), 493–503. https://doi.org/10.1007/s00418-003-0589-1.

Sakariassen, P.Ø., Prestegarden, L., Wang, J., Skaftnesmo, K.-O., Mahesparan, R., Molthoff, C., et al. (2006). Angiogenesis-independent tumor growth mediated by stem-like cancer cells. *Proceedings of the National Academy of Sciences of the United States of America, 103*(44), 16466–16471. https://doi.org/10.1073/pnas.0607668103.

Salmon, H., Franciszkiewicz, K., Damotte, D., Dieu-Nosjean, M. C., Validire, P., Trautmann, A., et al. (2012). Matrix architecture defines the preferential localization and migration of T cells into the stroma of human lung tumors. *Journal of Clinical Investigation, 122*(3), 899–910. https://doi.org/10.1172/JCI45817.

Sappino, A., Skalli, O., Jackson, B., Schürch, W., & Gabbiani, G. (1988). Smooth-muscle differentiation in stromal cells of malignant and non-malignant breast tissues. *International Journal of Cancer, 41*(5), 707–712. https://doi.org/10.1002/ijc.2910410512.

Sappino, A. P., Schurch, W., & Gabbiani, G. (1990). Differentiation repertoire of fibroblastic cells: Expression of cytoskeletal proteins as marker of phenotypic modulations. *Laboratory Investigation, 63*(2), 144–161.

Sato, T. N., Qin, Y., Kozak, C. A., & Audus, K. L. (1993). Tie-1 and tie-2 define another class of putative receptor tyrosine kinase genes expressed in early embryonic vascular system. *Proceedings of the National Academy of Sciences of the United States of America, 90*(20), 9355–9358. https://doi.org/10.1073/pnas.90.20.9355.

Schnurch, H., & Risau, W. (1993). Expression of tie-2, a member of a novel family of receptor tyrosine kinases, in the endothelial cell lineage. *Development, 119*(3), 957–968.

Shao, E. S., Lin, L., Yao, Y., & Boström, K. I. (2009). Expression of vascular endothelial growth factor is coordinately regulated by the activin-like kinase receptors 1 and 5 in endothelial cells. *Blood, 114*(10), 2197–2206. https://doi.org/10.1182/blood-2009-01-199166.

Shi, Q., Rafii, S., Wu, M. H., Wijelath, E. S., Yu, C., Ishida, A., et al. (1998). Evidence for circulating bone marrow-derived endothelial cells. *Blood, 92*(2), 362–367. https://www.ncbi.nlm.nih.gov/pubmed/9657732.

Shmelkov, S. V., St. Clair, R., Lyden, D., & Rafii, S. (2005). AC133/CD133/prominin-1. *International Journal of Biochemistry and Cell Biology, 37*(4), 715–719. https://doi.org/10.1016/j.biocel.2004.08.010.

Shojaei, F., Wu, X., Malik, A. K., Zhong, C., Baldwin, M. E., Schanz, S., et al. (2007). Tumor refractoriness to anti-VEGF treatment is mediated by CD11b +Gr1+ myeloid cells. *Nature Biotechnology, 25*(8), 911–920. https://doi.org/10.1038/nbt1323.

Sica, A., & Massarotti, M. (2017). Myeloid suppressor cells in cancer and autoimmunity. *Journal of Autoimmunity, 85*, 117–125. https://doi.org/10.1016/j.jaut.2017.07.010.

Singh, S. K., Clarke, I. D., Terasaki, M., Bonn, V. E., Hawkins, C., Squire, J., et al. (2003). Identification of a cancer stem cell in human brain tumors. *Cancer Research, 63*(18), 5821–5828.

Solovey, A., Lin, Y., Browne, P., Choong, S., Wayner, E., & Hebbel, R. P. (1997). Circulating activated endothelial cells in sickle cell anemia. *New England Journal of Medicine, 337*(22), 1584–1590. https://doi.org/10.1056/NEJM199711273372203.

Spira, A. I., & Carducci, M. A. (2003). Differentiation therapy. *Current Opinion in Pharmacology, 3*(4), 338–343. https://doi.org/10.1016/S1471-4892(03)00081-X.

Strehlow, K., Werner, N., Berweiler, J., Link, A., Dirnagl, U., Priller, J., et al. (2003). Estrogen increases bone marrow–derived endothelial progenitor cell production and diminishes neointima formation. *Circulation, 107*(24), 3059–3065. https://doi.org/10.1161/01.cir.0000077911.81151.30.

Stupp, R., & Hegi, M. E. (2007). Targeting brain-tumor stem cells. *Nature Biotechnology, 25*(2), 193–194. https://doi.org/10.1038/nbt0207-193.

Stylianopoulos, T., Munn, L. L., & Jain, R. K. (2018). Reengineering the physical microenvironment of tumors to improve drug delivery and efficacy: From mathematical modeling to bench to bedside. *Trends in Cancer, 4*(4), 292–319. https://doi.org/10.1016/j.trecan.2018.02.005.

Suda, T. (2000). Hematopoiesis and angiogenesis. *International Journal of Hematology, 71*(2), 99–107.

Takakura, N., Huang, X. L., Naruse, T., Hamaguchi, I., Dumont, D. J., Yancopoulos, G. D., et al. (1998). Critical role of the TIE2 endothelial cell receptor in the development of definitive hematopoiesis. *Immunity, 9*(5), 677–686. https://doi.org/10.1016/S1074-7613(00)80665-2.

Tanaka, Y., Abe, M., Hiasa, M., Oda, A., Amou, H., Nakano, A., et al. (2007). Myeloma cell-osteoclast interaction enhances angiogenesis together with bone resorption: A role for vascular endothelial cell growth factor and osteopontin. *Clinical Cancer Research, 13*(3), 816–823. https://doi.org/10.1158/1078-0432.ccr-06-2258.

Tavor, S. (2004). CXCR4 regulates migration and development of human acute myelogenous leukemia stem cells in transplanted NOD/SCID mice. *Cancer Research, 64*(8), 2817–2824. https://doi.org/10.1158/0008-5472.can-03-3693.

Vajkoczy, P., Blum, S., Lamparter, M., Mailhammer, R., Erber, R., Engelhardt, B., et al. (2003). Multistep nature of microvascular recruitment of ex vivo-expanded embryonic endothelial progenitor cells during tumor angiogenesis. *Journal of Experimental Medicine, 197*(12), 1755–1765. https://doi.org/10.1084/jem.20021659.

Vasa, M., Fichtlscherer, S., Aicher, A., Adler, K., Urbich, C., Martin, H., et al. (2001). Number and migratory activity of circulating endothelial progenitor cells inversely correlate with risk factors for coronary artery disease. *Circulation Research, 89*(1), E1–7.

Venneri, M. A., De Palma, M., Ponzoni, M., Pucci, F., Scielzo, C., Zonari, E., et al. (2007). Identification of proangiogenic TIE2-expressing monocytes (TEMs) in human peripheral blood and cancer. *Blood, 109*(12), 5276–5285. https://doi.org/10.1182/blood-2006-10-053504.

Vittet, D., Prandini, M. H., Berthier, R., Schweitzer, A., Martin-Sisteron, H., Uzan, G., et al. (1996). Embryonic stem cells differentiate in vitro to endothelial cells through successive maturation steps. *Blood, 88*(9), 3424–3431.

Wang, R., Chadalavada, K., Wilshire, J., Kowalik, U., Hovinga, K. E., Geber, A., et al. (2010). Glioblastoma stem-like cells give rise to tumour endothelium. *Nature, 468*(7325), 829–833. https://doi.org/10.1038/nature09624.

Welford, A. F., Biziato, D., Coffelt, S. B., Nucera, S., Fisher, M., Pucci, F., et al. (2011). TIE2-expressing macrophages limit the therapeutic efficacy of the vascular-disrupting agent combretastatin A4 phosphate in mice. *The Journal of Clinical Investigation, 121*(5), 1969–1973. https://doi.org/10.1172/JCI44562.

Wicha, M. S., Liu, S., & Dontu, G. (2006). Cancer stem cells: An old idea - A paradigm shift. *Cancer Research, 66*(4), 1883–1890. https://doi.org/10.1158/0008-5472.CAN-05-3153.

Yin, A. H., Miraglia, S., Zanjani, E. D., Almeida-Porada, G., Ogawa, M., Leary, A. G., et al. (1997). AC133, a novel marker for human hematopoietic stem and progenitor cells. *Blood, 90*(12), 5002–5012.

Yin, T., & Li, L. (2006). The stem cell niches in bone. *Journal of Clinical Investigation, 116*(5), 1195–1201. https://doi.org/10.1172/JCI28568.

Zeisberg, E. M., Potenta, S., Xie, L., Zeisberg, M., & Kalluri, R. (2007). Discovery of endothelial to mesenchymal transition as a source for carcinoma-associated fibroblasts. *Cancer Research, 67*(21), 10123–10128. https://doi.org/10.1158/0008-5472.CAN-07-3127.

CHAPTER 11

Therapeutic strategies

11.1 Anti-inflammatory compounds

Long-term use of various anti-inflammatory compounds, e.g., aspirin, nonsteroidal anti-inflammatory drugs (NSAIDs), and cyclooxygenase-2 (COX-2) inhibitors, reduces colon cancer risk by 50%, gastric and esophageal cancer risk by 40 per cent, and breast cancer risk by 20 per cent (Shacter and Weitzman, 2002), confirming a promoting role for inflammation in tumor progression. COX-2 promotes tumor progression by inducing epithelial—mesenchymal transition (EMT) and stimulating vascular endothelial growth factor (VEGF)-mediated angiogenesis and matrix metalloproteinase-14 (MMP-14)-driven invasion (Giannoni et al., 2010). In this respect, preclinical and clinical trials demonstrated that COX-2 inhibitors such as celecoxib (Celebrex) represent a promising strategy in the prevention and treatment of solid tumors (Ghosh et al., 2010).

In addition to influencing proliferation and survival of neoplastic cells, tumor-infiltrating inflammatory cells are also involved in the regulation of tumor angiogenesis. Inflammatory lesions preceding early stages of tumor progression contribute to create an environment favorable to tumor progression. For example, Crohn's disease or ulcerative colitis, or chronic reflux esophagitis in Barrett syndrome are associated with increased risk of developing colorectal or esophageal cancer, respectively (van der Woude et al., 2004). *Helicobacter pylori* may produce inflammatory toxins, impair antitumor responses by the immune system, or enforce signaling pathways involved in tumor development (Javed et al., 2019).

Furthermore, *Helicobacter pylori* induced gastritis is associated with gastric carcinoma and mucosa-associated lymphoid tissue B-cell lymphoma (MALT lymphoma) (Mazzucchelli et al., 1999). In most of these cases, carcinogenesis is believed to be secondary to the creation of a local chronic inflammatory state. The microbiota has been involved in the development of different tumor types and impacts host response to chemotherapy and immunotherapies (Dzutsev et al., 2014). The characterization of stromal components is

Tumor Microenvironment Regulation of Tumor Expansion
ISBN 978-0-12-822803-6
https://doi.org/10.1016/B978-0-12-822803-6.00007-7

important to allow the development of new therapeutic strategies (Taube et al., 2012), including inhibitors of cancer associated fibroblast (CAF) activation and proliferation in pancreatic cancer (Ishii et al., 2019; Whittle & Hingorani, 2019) and inhibitors of colony stimulating factor-1 (CSF-1) and myeloid derived stem cells (MDSCs) (Schupp et al., 2019).

11.2 Therapeutic vaccines and immune checkpoints

The term immunotherapy was introduced by Paul Ehrlich in 1909. Specific antibodies would, like "magic bullets," seek out pathogenetic targets and, by suitable modification, a "toxophore" group would be added to the antibodies to kill the target organism (Note 1). Clinical trials with tumor antigen vaccines, systemic administration of cytokines, adaptive transfer of T cells specific for tumor antigens or antibodies demonstrate that the immune system is capable of rejecting established tumors. Therapeutic vaccines promoting immune targeting of tumor vascular cells limit tumor-associated angiogenesis and disease progression, through induction of antigen-specific T $CD8^+$ cell response. However, targeting antigens by peptide vaccination attacking angiogenic targets have shown only limited therapeutic beneficial results (Rahat, 2019).

Vaccination based on tumor blood vessel associated antigens (TBVAs) normalizes the vasculature, induces T cell effector response, and inhibits tumor growth in murine models (Ishizaki et al., 2006). Two major types of vaccines are being developed, namely vaccines against defined angiogenesis-associated antigens, and whole endothelial cell vaccines, expressing numerous angiogenic antigens, using whole endothelial cells or isolated proteins from endothelial cell membranes (Wagner et al., 2015). Antigens present in early neoplastic tissues are transported to lymphoid organs by dendritic cells that activate adaptive immune responses, e.g., B and T lymphocytes, resulting in chronic activation of innate immune cells in neoplastic tissues. In some cases, the presence of tumor infiltrating lymphocytes (TILs) in primary tumors is associated with a better prognosis (Yu and Fu, 2006). Failure to appropriately activate adaptive immune cells results in failed recruitment of innate immune cells into the tumor microenvironment (De Visser et al., 2005). The initiation of the adaptive immune response depends on antigen presentation between the MHC and the T-cell receptor (signal 1), and via CD28 costimulation with B7 (CD80 or CD86) (signal 2).

The duration and amplitude of the response is modulated by different checkpoints (Table 11.1). One of these is cytotoxic T-lymphocyte-associated

Table 11.1 Inflammatory cells expressing checkpoint inhibitors.

B, NK (PD1-l)
Tregs (CTLA-4)
CD8T, CD4T, NK, Tregs, Dendritic cells, Monocytes (LAG-3[a])
CD8T, CD4T, Dendritic cells, Monocytes (TIM-3[b])
T, Neutrophils, Monocytes (VISTA[c])

[a]Lymphocyte activation gene-3.
[b]T cell immunoglobulin and mucin domain 3.
[c]V domain containing immunoglobulin suppressor of T cell activation.

protein-4 (CTLA-4) (Note 2) which has a much stronger affinity for the B7 molecule than CD28 and leads to attenuation of the immune response. Another one is the programmed death 1 (PD-1)/programmed death ligand-1 (PD-L1) pathway (Note 3). Upregulation of immune checkpoint receptors occurs during T cell activation in an effort to prevent damage from excessive immune response. Tumors can coopt this pathway and dampen the local host immune response to the tumor. The expression of immune checkpoint gene, such as PD-1 and PD-L1 and CTLA-4, plays an important role in downregulation of the immune response to a tumor by preventing the activation of T-cells. PD-L1-mediated adaptive immune resistance was first described in melanoma (Taube et al., 2012). Checkpoint inhibitor immunotherapies work by blocking the immune inhibitors CTLA4 or PD-1/PD-L1, allowing natural host antitumor immune response. Many of the clinical successes with these agents have been obtained in patients with melanoma (Hamid et al., 2013; Postow et al., 2015). Ipilimumab, a fully human IgG which binds and blocks CTLA-4 interactions with its ligands CD80 and CD86, was the first checkpoint inhibitor agent approved by the Food and Drug Administration (FDA) for the treatment of advanced metastatic melanoma (Hodi et al., 2015). The FDA has also approved nivolumab and pembrolizumab, two antibodies that target PD-1 and block its interactions with PDL-1 and PDL-2, for the treatment of other tumors, including squamous and nonsquamous non—small-cell lung carcinoma, bladder carcinoma, Hodgkin lymphoma, head and neck squamous cell carcinoma, Merkel cell carcinoma, and advanced clear cell renal cell carcinoma (Gunturi & McDermott, 2015; Khoja, Butler, & Kang, 2015).

Immunotherapeutic agents may be combined for their synergistic effect as it has been demonstrated in preclinical study conducted on murine tumor models (Curran et al., 2010). PD-L1 expression on cancer cells as a biomarker for patient selection has been approved for patients who are

PD-L1 tumor cell positive (Topalian et al., 2016). PD-L1 expression on immune cells, including dendritic cells and macrophages, in tumor microenvironment can determine the efficacy of PD-L1 pathway blockade-mediated tumor regression (Lin et al., 2018). PDL-1 expression is associated with poorer prognosis in a variety of solid tumors, including melanoma, renal, and lung cancer (Wang, Liu, & Liu, 2017; Wu, Wu, & Li, 2015). The expression of PDL-1 on human endothelial cells was associated with reduced T-cell responses (LaGier and Pober, 2006). VEGF-A, produced by both tumor and endothelial cells, is involved in the creation of an immunosuppressive microenvironment through enhancing the production of PD-1 (Oron et al., 2015). The side effects associated with checkpoint inhibition are termed immune-related adverse effects. The combination of anti-CTLA-4 and anti PD-1 antibodies had a much greater response rate compared to the use of either antibody alone, even though it also resulted in greater side effects and the possibility of relapse. The use of engineered T cells equipped with chimeric antigen receptors with predefined specificity, referred as CAR T cells, have shown positive activity against B cell human malignancies. More recently, it has been proposed that tumor vasculature may be considered as an attractive CAR T target in solid tumors (Akbari et al., 2019).

11.3 Targeting the tumor vasculature

Vascular cells are more readily accessible to circulating effector cells, antibodies, and other therapeutic molecules than tumor cells. When compared to tumor cells, tumor-associated vascular cells are genetically more stable and exhibit more consistent expression of MHC class I and II molecules required for specific T cell recognition. Angiogenesis inhibitors are classified into direct and indirect agents. Direct endogenous inhibitors target vascular endothelial cells, while indirect inhibitors target tumor cells or tumor-associated stromal cells. Many antiangiogenic agents create transient state of vascular normalization in the tumor microenvironment, during which they are more receptive to delivery chemotherapeutic drugs, by increasing blood flow.

VEGF blockade induces vascular normalization by pruning immature blood vessels and inducing pericyte recruitment (Winkler et al., 2004). Excessive pruning leads to an insufficient vasculature and consequent hypoxia and shorter normalization window (Jain, 2014). Studies with MRI provide some evidence of tumor vessel normalization after treatment with anti-VEGF agents (Batchelor et al., 2007). The normalization window closes when vessels are excessively pruned or when the tumor uses other angiogenesis pathways or methods of vessel recruitment. Cytotoxic therapy given during the normalization window has a better outcome than the cytotoxic therapy given before or after the window (Winkler et al., 2004). Tumor vascular normalization is accompanied by increased pericyte coverage, while pericyte deficiency could be partly responsible for vessel abnormalities in tumor blood vessels (Gerhardt and Semb, 2008) and partial dissociation of pericytes contributes to increased tumor vascular permeability (Hashizume et al., 2000). Pericyte coverage promotes resistance through direct support or paracrine interactions with endothelial cells, and tumor vessels covered by pericytes are less sensitive to VEGF blockade (Ribatti et al., 2011). Normalization can slow down tumor growth and restores structural and functional integrity of tumor vasculature, with enhanced pericyte coverage, homogeneous basement membrane, enhanced perfusion, and reduced hypoxia.

The state of vascular normalization generally is transient, and in this context tumors are more receptive to delivery of chemotherapeutic drug and immune cell populations (Griffioen, 2008). However, there appears to be a high degree of variation in a temporal window of vascular normalization for various antiangiogenic agents (Hernández-Agudo et al., 2016). Vascular normalization can convert the immunosuppressive tumor microenvironment into an immune-stimulatory one by promoting the antitumor activity of immune effector cells (Fukumura et al., 2018). Metronomic chemotherapy (Note 4) has also been shown to induce tumor vessel normalization (Romiti et al., 2013). The window of vascular normalization occurs transiently during the first days of treatment. Moreover, while normalization is effective in the early stages of tumor development, no apparent differences in tumor growth are observed when it is applied in later stages (Goel et al., 2008). When the tumor blood vessels lose their maturation (Peng et al., 2012), hypoxia reincreases and prolonged VEGF inhibition further increases local hypoxia (Winkler et al., 2004), which in turn induces systemic secretion of other angiogenic cytokines (Casanovas, Hicklin, Bergers, & Hanahan, 2005; Ebos et al., 2009). Hypoxia induces upregulation of VEGF expression through hypoxia inducible factor 1 alpha (HIF1-α) (Forsythe et al., 1996).

Hypoxia within tumor increases during treatment with an antiangiogenic agent, inducing pH drop and consequent acidosis (Jain, 2014). Hypoxic conditions could also select more malignant cells, i.e., those able to grow in hypoxic conditions (Wilson and Hay, 2011). It produces a

Table 11.2 Effects of inhibiting VEGF signaling on different conponents of tumor blood vessels.

Endothelial cells
Baseline conditions in tumors (abnormal; abnormal gene expression; leakage; growth factors dependency) Inhibition of VEGF signaling (loss of sprouts; loss of fenestrations; reduction of VEGFR-2 expression; loss of lumen; apoptosis; regression)
Pericytes
Baseline conditions in tumors (present but abnormal; loose association with endothelial cells) Inhibition of VEGF signaling (change in phenotype; normalization of endothelial cell association)
Basement membrane
Baseline conditions in tumors (present but abnormal; loose association with endothelial cells) Inhibition of VEGF signaling (persistent after regression of endothelial cells)

pressure mechanism that selects tumor cells with increased aggressiveness and lower sensitivity to antiangiogenic therapy (Ebos et al., 2009). In fact, blood supply to the tumor remains heterogeneous and hypoxia increases while the permeability of tumor blood vessels is reduced, favoring disappointing results of clinical trials where antiangiogenic therapies have been combined with systemic delivery of chemotherapeutic agents. Hypoxia exerts a direct positive effect on c-MET and phospho-c-met expression, and the activation of the hepatocyte growth factor (HGF)/c-MET pathway promotes tumor invasion and invasiveness (Pennacchietti et al., 2003; Trusolino, Bertotti, & Comoglio, 2010). Upon interaction with c-MET, integrins enhance tumor cell invasiveness (Jahangiri and Chandra, 2014). Given the major role of VEGF in angiogenesis, an array of antiangiogenic therapies has been developed to antagonize the VEGF—VEGFR axis (Table 11.2). However, despite the growing list of FDA approvals, antiangiogenic drugs had a modest impact on overall patient survival.

The discrepancy with the results obtained in preclinical trials is likely to be due, in part, to preclinical models where agents were administered to small, fast growing, and highly angiogenic tumors, whereas their efficacy in clinical trials has almost exclusively been tested in late stages, in well-established cancers, well vascularized, and less likely to be dependent upon new vessel formation. Bevacizumab (Avastin, Genentech), a humanized anti-VEGF-A monoclonal antibody (Presta et al., 1997), was the first antiangiogenic drug approved by the FDA as first- or second-line therapy in refractory patients in the setting of metastatic colorectal cancer, non—small-cell lung cancer (NSCLC), recurrent gliobastoma, and metastatic renal cell carcinoma, usually in combination with chemotherapeutic agents (Hurwitz et al., 2004; Sandler et al., 2006). The combination of bevacizumab and chemotherapy increases the delivery of the chemotherapeutic agent to the cancer tissue by increasing blood flow (Jain, 2014). Large randomized phase 3 clinical trials of bevacizumab therapy in combination with systemic chemotherapy have demonstrated significant improvements in progression-free survival and overall survival when compared with systemic chemotherapy alone.

Multiple clinical trials in the first- and second-line treatment of metastatic colorectal cancer have all confirmed that the addition of bevacizumab to standard first-line chemotherapy regimens significantly improves patient outcomes. However, in response to chemotherapy, tumor cells can themselves secrete factors that result in their enhanced resistance. Aflibercept is a soluble VEGF decoy receptor that consists of the extracellular domains of VEGFRs 1 and 2 and of the Fc portion of human IgG1. Owing to its structure, aflibercept can neutralize both VEGF and placental growth factor (PlGF) (Ciombor et al., 2013), and its use in patient-derived xenografts resulted in higher tumor suppressive activity, compared with bevacizumab (Chiron et al., 2014). Ramucirumab (Cyramza) is a humanized monoclonal antibody that selectively targets VEGFR-2 and blocks the interaction of VEGFR-2 with VEGF. It has been approved for application in colorectal cancer, NSCLC, and stomach adenocarcinoma (Aprile et al., 2014).

Tyrosine kinase inhibitors (TKIs) are small molecules that interrupt ATP binding to the tyrosine kinase catalytic domain through competition, thereby inhibiting the downstream receptor signaling. TKIs, including sunitinib (Sutent), sorafenib (Nexavar), everolimus (Afinitor), temsirolimus (Torisel), and pazopanib (Votrient), interfere with VEGFR-1 and VEGFR-2, PDGFR, fibroblast growth factor receptors (FGFRs), and Tie signaling, and have been approved for the treatment of patients with advanced cancer (Table 11.3). The impact of these treatments on overall survival was limited. Multiple mechanisms of drug-induced resistance against angiogenesis inhibitors exist. The presence of anti-VEGF resistant cancers suggest alternative angiogenic pathways, including angiopoietin-1 and angiopoietin-2 (Ang-1 and Ang-2), epidermal growth factor (EGF), FGF, granulocyte colony stimulating factor (G-CSF), HGF, insulin like growth factor (ILGF), PDGF,

Table 11.3 Approved tyrosine kinase inhibitors for cancer therapy and their indications.

Sorafenib (renal cell carcinoma; hepatocellular carcinoma; thyroid cancer
Sunitinib (renal cell carcinoma, pancreatic neuroendocrine tumors, gastrointestinal stromal tumors)
Pazopanib (renal cell carcinoma; soft tissue sarcoma)
Axitinib (renal cell carcinoma)
Vandetanib (medullary thyroid cancer)
Cabozantinib (medullary thyroid cancer; renal cell carcinoma)
Lenvatinib (thyroid cancer; renal cell carcinoma)

stromal derived factor-1 (SDF-1), and transforming growth factor (TGF). Targeting multiple growth factors simultaneously or sequentially would be a successful approach to overcome resistance. Blockade of VEGF and Ang-2 in preclinical studies suppressed revascularization and tumor progression of cancer resistant to anti-VEGF therapy (Rigamonti et al., 2014). Treatment of RIP1-Tag2 pancreatic neuroendocrine tumors with combined Ang-2 and VEGFR-2 blockers decreased infiltration of Tie-2 monocytes and suppressed revascularization and tumor progression (Rigamonti et al., 2014). Combining the FGFR inhibitor PD173074 with bevacizumab in xenografted mouse models with head and neck squamous cell carcinoma completely abolished tumor growth (Gyanchandani et al., 2013). Treatment of patients with recurrent and persistent endometrial cancer with the dual VEGFR/FGFR inhibitor, brivanib, improved their progression-free survival (Powell et al., 2014). In transgenic mouse models, treatment with a combination of two anti-PDGFR agents, imatinib and SU11248, cyclophosphamide, and/or an anti-VEGFR agent, significantly improved antitumor responses (Pietras and Hanahan, 2005).

Some antiangiogenic molecules are more capable to induction of resistance than others, and this event is responsible of the discontinuation of the therapy. Antiangiogenic therapy can improve anti-PDL-1 treatment in animal models, and treatment with a bispecific anti-VEGF/Ang-2 antibody potentiated the activity of anti-PDL-1 treatment in multiple models. The successful development of anti-VEGF targeted therapy will require a greater understanding of how tumors become vascularized and how they evade the effects of antiangiogenic therapy. VEGF blockade aggravates tumor hypoxia, which upregulates the production of other angiogenic factors in the tumor microenvironment, including VEGF, PlGF, FGF-2, and PDGF. Moreover, HIF1 inhibits the production of antiangiogenic factors. Tumor hypoxia induces nutrient deprivation that activates the tumor nonoxidative metabolism (glycolysis) contributing to the increase of resistant tumor colonies (Quintieri et al., 2014). Several genes involved in glycolysis are under HIF1 control, including GLUT1, GLUT3, PDK1, and LDHA. Tumor hypoxia contributes to the immunosuppressive effect of the tumor microenvironment, leading to PDL-1 upregulation (Noman et al., 2014), enhancing the activity of suppressor T regulatory cells and inhibiting effector T cell functions (Westendorf et al., 2017). Interleukin 8 (IL-8) is involved in anti-VEGF tumor resistance in sunitinib-treated renal carcinoma (Huang et al., 2010). IL-17 mobilizes the G-CSF-dependent recruitment of $CD11b^{+}$ $Gr1^{+}$ immature myeloid cells, involved in anti-VEGF tumor refractoriness (Chung et al., 2013). Human late-stage breast cancers expressed several angiogenic cytokines in contrast to earlier stage lesions, which expressed only VEGF (Relf et al., 1997).

Vasculogenic mimicry might serve as an escape mechanism from antiangiogenic therapies (Note 5). Also, vessel cooption might represent an intrinsic resistance mechanism to antiangiogenic therapies (Note 6). Targeting VEGF and other pathways implicated in angiogenesis should result in more effective tumor growth inhibition. Moreover, rational combinations of antiangiogenic agents based on pharmacokinetic and pharmacodynamics data are needed to overcome resistance, and it is extremely important to determine the optimal duration and scheduling of anti-VEGF agents. In addition to using antiangiogenic agents that target more than one proangiogenic factor, another strategy is to use combined modalities. It is necessary to select drugs with a low resistance to prevent or overcome the development of resistance, and combine different molecules to improve the therapeutic effect (Ribatti, 2016). The most popular combinatorial treatment strategy to date is the coadministration of angiogenesis inhibitors with chemotherapy. Combining immunotherapy with antiangiogenic therapy had reciprocal beneficial effects: antiangiogenic drugs seem to block the negative immune signals by increasing ratio of anti/protumor immune cells and decreasing immune checkpoints expression, while immunotherapy seems to restore an immune-supportive microenvironment and to promote vascular normalization increasing lymphocyte infiltration.

11.4 Notes

1. In 1909, Paul Ehrlich formulated the hypothesis that host defense may prevent neoplastic cells from developing into tumors. He stated that: "in the enormously complicated course of fetal and post-fetal development,

aberrant cells become unusually common. Fortunately, in the majority of people, they remain completely latent thanks to the organism's positive mechanisms." The first clear demonstration of specific capability to stimulate an immune response was made by Gross in 1953 after intradermal immunization of C3H mice, obtained by continuous brother to sister mating for more than 20 years, against a sarcoma. In 1957, Sir Frank Macfarlane Burnet hypothesized that tumor cell neoantigens induce an immunological reaction against cancer and subsequently formulated the immune surveillance theory. He wrote that: "It is by no means inconceivable that small accumulation of tumor cells may develop and because of their possession of new antigenic potentialities provoke an effective immunological reaction with regression of the tumor and no clinical hint of its existence" (Burnet, 1970).

2. CTLA4 (CD 152) is a membrane glycoprotein expressed by Tregs that inhibits early T cell activation and plays an important role in the priming phase of the immune response (Teft, Kirchhof, & Madrenas, 2006). CTLA-4 was initially studied as a homolog of CD28 but differed in that its ligation suppressed IL-2 production, whereas CD28 promoted IL-2 biosynthesis when T cells were activated via the TCR (Krummel and Allison, 1995).
3. PD-1 was first described as an inhibitory receptor on T cells that suppressed signaling by the TCR (Okazaki & Honjo, 2007). PD-1 blocks T cell activation and is associated with chronically activated and exhausted T cells, such as those found in the tumor microenvironment (Keir et al., 2008).
4. Metronomic chemotherapy schedules include a more continuous administration of conventional cytotoxic agents at lower doses than classical chemotherapy, whereby the cumulative dose over time often exceeds that of conventional 3-week regimens. Metronomic chemotherapy can be administered as a primary treatment, as a maintenance therapy after conventional chemotherapy and has been shown to be a valid treatment option in different malignant tumors.
5. In 1999, (Maniotis et al., 1999) described a new model of formation of vascular channels by human melanoma cells and called it "vasculogenic mimicry" to emphasize the de novo generation of blood vessels without the participation of endothelial cell and independent of angiogenesis. Microarray gene chip analysis of a highly aggressive compared with poorly aggressive human cutaneous melanoma cell lines revealed a significant increase in the expression of laminin 5 and MMP-1,

MMP-2, MMP-9, and MT1-MMP in the highly aggressive cells, suggesting that they interact with and alter their extracellular environment differently than the poorly aggressive cells, and that increased expression of MMP-2 and MT1-MMP along with matrix deposition of laminin 5 are required for their mimicry. Vasculogenic mimicry can serve as a marker for tumor metastasis, a poor prognosis, worse survival, and a highest risk of cancer recurrence.

6. In 1999, (Holash et al., 1999) reported that tumor cells migrate to host organ blood vessels in sites of metastases, or in vascularized organs such as the brain, and initiate blood-vessel-dependent tumor growth as opposed to classic angiogenesis. These vessels then regress owing to apoptosis of the constituent endothelial cell, apparently mediated by Ang-2. Lastly, at the periphery of the growing tumor mass angiogenesis occurs by cooperative interaction of VEGF and Ang-2. The ability of cancer cells to coopt the vessels initiates the process though few of them will intravasate, starting the metastatic process. Treatment of glioma with a monoclonal antibody anti-VEGFR-2 induces cooption of quiescent cerebral vessels, and treatment of cerebral melanoma metastasis with the TKIZD6474 is associated with increase in vessel cooption.

References

Akbari, P., Huijbers, E. J. M., Themeli, M., Griffioen, A. W., & van Beijnum, J. R. (2019). The tumor vasculature an attractive CAR T cell target in solid tumors. *Angiogenesis, 22*(4), 473–475. https://doi.org/10.1007/s10456-019-09687-9.

Aprile, G., Rijavec, E., Fontanella, C., Rihawi, K., & Grossi, F. (2014). Ramucirumab: Preclinical research and clinical development. *OncoTargets and Therapy,* 7, 1997–2006. https://doi.org/10.2147/OTT.S61132.

Batchelor, T. T., Sorensen, A. G., di Tomaso, E., Zhang, W.-T., Duda, D. G., Cohen, K. S., et al. (2007). AZD2171, a pan-VEGF receptor tyrosine kinase inhibitor, normalizes tumor vasculature and alleviates edema in glioblastoma patients. *Cancer Cell, 11*(1), 83–95. https://doi.org/10.1016/j.ccr.2006.11.021.

Burnet, M. (1970). In *Immunological surveillance* (pp. 161–185). Elsevier. https://doi.org/10.1016/b978-0-08-017481-5.50012-2.

Casanovas, O., Hicklin, D. J., Bergers, G., & Hanahan, D. (2005). Drug resistance by evasion of antiangiogenic targeting of VEGF signaling in late-stage pancreatic islet tumors. *Cancer Cell, 8*(4), 299–309. https://doi.org/10.1016/j.ccr.2005.09.005.

Chiron, M., Bagley, R. G., Pollard, J., Mankoo, P. K., Henry, C., Vincent, L., et al. (2014). Differential antitumor activity of aflibercept and bevacizumab in patient-derived xenograft models of colorectal cancer. *Molecular Cancer Therapeutics, 13*(6), 1636–1644. https://doi.org/10.1158/1535-7163.MCT-13-0753.

Chung, A. S., Wu, X., Zhuang, G., Ngu, H., Kasman, I., Zhang, J., et al. (2013). An interleukin-17–mediated paracrine network promotes tumor resistance to anti-angiogenic therapy. *Nature Medicine, 19*(9), 1114–1123. https://doi.org/10.1038/nm.3291.

Ciombor, K. K., Berlin, J., & Chan, E. (2013). Aflibercept. *Clinical Cancer Research, 19*(8), 1920–1925. https://doi.org/10.1158/1078-0432.CCR-12-2911.

Curran, M. A., Montalvo, W., Yagita, H., & Allison, J. P. (2010). PD-1 and CTLA-4 combination blockade expands infiltrating T cells and reduces regulatory T and myeloid cells within B16 melanoma tumors. *Proceedings of the National Academy of Sciences of the United States of America, 107*(9), 4275–4280. https://doi.org/10.1073/pnas.0915174107.

De Visser, K. E., Korets, L. V., & Coussens, L. M. (2005). De novo carcinogenesis promoted by chronic inflammation is B lymphocyte dependent. *Cancer Cell,* 7(5), 411–423. https://doi.org/10.1016/j.ccr.2005.04.014.

Dzutsev, A., Goldszmid, R. S., Viaud, S., Zitvogel, L., & Trinchieri, G. (2014). The role of the microbiota in inflammation, carcinogenesis, and cancer therapy. *European Journal of Immunology, 45*(1), 17–31. https://doi.org/10.1002/eji.201444972.

Ebos, J. M. L., Lee, C. R., Cruz-Munoz, W., Bjarnason, G. A., Christensen, J. G., & Kerbel, R. S. (2009). Accelerated metastasis after short-term treatment with a potent inhibitor of tumor angiogenesis. *Cancer Cell, 15*(3), 232–239. https://doi.org/10.1016/j.ccr.2009.01.021.

Forsythe, J. A., Jiang, B. H., Iyer, N. V., Agani, F., Leung, S. W., Koos, R. D., et al. (1996). Activation of vascular endothelial growth factor gene transcription by hypoxia-inducible factor 1. *Molecular and Cellular Biology, 16*(9), 4604–4613. https://doi.org/10.1128/MCB.16.9.4604.

Fukumura, D., Kloepper, J., Amoozgar, Z., Duda, D. G., & Jain, R. K. (2018). Enhancing cancer immunotherapy using antiangiogenics: Opportunities and challenges. *Nature Reviews Clinical Oncology, 15*(5), 325–340. https://doi.org/10.1038/nrclinonc.2018.29.

Gerhardt, H., & Semb, H. (2008). Pericytes: Gatekeepers in tumour cell metastasis? *Journal of Molecular Medicine, 86*(2), 135–144. https://doi.org/10.1007/s00109-007-0258-2.

Ghosh, N., Chaki, R., Mandal, V., & Mandal, S. C. (2010). Cox-2 as a target for cancer chemotherapy. *Pharmacological Reports, 62*(2), 233–244. https://doi.org/10.1016/S1734-1140(10)70262-0.

Giannoni, E., Bianchini, F., Masieri, L., Serni, S., Torre, E., Calorini, L., et al. (2010). Reciprocal activation of prostate cancer cells and cancer-associated fibroblasts stimulates epithelial-mesenchymal transition and cancer stemness. *Cancer Research, 70*(17), 6945–6956. https://doi.org/10.1158/0008-5472.CAN-10-0785.

Goel, Wong, A., & Jain. (2008). Vascular normalization as a therapeutic strategy for malignant and non-malignant disease. *Cold Spring Harbor Perspectives in Medicine, 2,* a006486. Griffioen AW. Cancer Immunol Immunoter, 57, 1553–1558.

Griffioen, A. W. (2008). Anti-angiogenesis: Making the tumor vulnerable to the immune system. *Cancer Immunology, Immunotherapy, 57*(10), 1553–1558. https://doi.org/10.1007/s00262-008-0524-3.

Gunturi, A., & McDermott, D. F. (2015). Nivolumab for the treatment of cancer. *Expert Opinion on Investigational Drugs, 24*(2), 253–260. https://doi.org/10.1517/13543784.2015.991819.

Gyanchandani, R., Ortega Alves, M. V., Myers, J. N., & Kim, S. (2013). A proangiogenic signature is revealed in FGF-mediated bevacizumab-resistant head and neck squamous cell carcinoma. *Molecular Cancer Research, 11*(12), 1585–1596. https://doi.org/10.1158/1541-7786.MCR-13-0358.

Hamid, O., Robert, C., Daud, A., Hodi, F. S., Hwu, W.-J., Kefford, R., et al. (2013). Safety and tumor responses with lambrolizumab (anti-PD-1) in melanoma. *The New England Journal of Medicine, 369*(2), 134–144. https://doi.org/10.1056/NEJMoa1305133.

Hashizume, H., Baluk, P., Morikawa, S., McLean, J. W., Thurston, G., Roberge, S., et al. (2000). Openings between defective endothelial cells explain tumor vessel leakiness. *American Journal of Pathology, 156*(4), 1363–1380. https://doi.org/10.1016/S0002-9440(10)65006-7.

Hernández-Agudo, E., Mondejar, T., Soto-Montenegro, M. L., Megías, D., Mouron, S., Sanchez, J., et al. (2016). Monitoring vascular normalization induced by antiangiogenic treatment with (18)F-fluoromisonidazole-PET. *Molecular Oncology, 10*(5), 704–718. https://doi.org/10.1016/j.molonc.2015.12.011.

Hodi, F. S., Ernstoff, M. S., Minor, D., Salama, A. K., Taylor, M., Ott, P. A., et al. (2015). Nivolumab and ipilimumab versus ipilimumab in untreated melanoma. *New England Journal of Medicine, 372*(21), 2006–2017. https://doi.org/10.1056/NEJMoa1414428.

Holash, J., Maisonpierre, P. C., Compton, D., Boland, P., Alexander, C. R., & Zagzag, D. (1999). Vessel cooption, regression and growth in tumors mediated by angiopoietins and VEGF. *Science, 28*, 1994–1998.

Huang, D., Ding, Y., Zhou, M., Rini, B. I., Petillo, D., Qian, C. N., et al. (2010). Interleukin-8 mediates resistance to antiangiogenic agent sunitinib in renal cell carcinoma. *Cancer Research, 70*(3), 1063–1071. https://doi.org/10.1158/0008-5472.CAN-09-3965.

Hurwitz, H., Fehrenbacher, L., Novotny, W., Cartwright, T., Hainsworth, J., Heim, W., et al. (2004). Bevacizumab plus irinotecan, fluoruracil, and leucovirin for metastatic colorectal cancer. *New England Journal of Medicine, 350*, 2335–2342.

Ishii, N., Araki, K., Yokobori, T., Hagiwara, K., Gantumur, D., Yamanaka, T., et al. (2019). Conophylline suppresses pancreatic cancer desmoplasia and cancer-promoting cytokines produced by cancer-associated fibroblasts. *Cancer Science, 110*(1), 334–344. https://doi.org/10.1111/cas.13847.

Ishizaki, H., Tsunoda, T., Wada, S., Yamauchi, M., Shibuya, M., & Tahara, H. (2006). Inhibition of tumor growth with antiangiogenic cancer vaccine using epitope peptides derived from human vascular endothelial growth factor receptor 1. *Clinical Cancer Research, 12*(19), 5841–5849. https://doi.org/10.1158/1078-0432.CCR-06-0750.

Jahangiri, N., & Chandra, A. (2014). Cross-activating c-Met/beta 1 integrin complex drives metastasis and invasive resistance in cancer. *Proceedings of the National Academy of Sciences of the United States of America, 114*, 605–622.

Jain, R. K. (2014). Antiangiogenesis strategies revisited: From starving tumors to alleviating hypoxia. *Cancer Cell, 26*(5), 605–622. https://doi.org/10.1016/j.ccell.2014.10.006.

Javed, S., Skoog, E. C., & Solnick, J. V. (2019). Impact of helicobacter pylori virulence factors on the host immune response and gastric pathology. In *Current Topics in Microbiology and Immunology, 421* pp. 21–52). Springer Verlag. https://doi.org/10.1007/978-3-030-15138-6_2.

Keir, M. E., Butte, M. J., Freeman, G. J., & Sharpe, A. H. (2008). PD-1 and its ligands in tolerance and immunity. *Annual Review of Immunology, 26*, 677–704. https://doi.org/10.1146/annurev.immunol.26.021607.090331.

Khoja, L., Butler, M. O., Kang, S. P., Ebbinghaus, S., & Joshua, A. M. (2015). Pembrolizumab. *Journal for Immunotherapy of Cancer, 18*(3), 36. https://doi.org/10.1186/s40425-015-0078-9. eCollection 2015.

Krummel, M. F., & Allison, J. P. (1995). CD28 and CTLA-4 have opposing effects on the response of T ceils to stimulation. *Journal of Experimental Medicine, 182*(2), 459–465. https://doi.org/10.1084/jem.182.2.459.

LaGier, A. J., & Pober, J. S. (2006). Immune accessory functions of human endothelial cells are modulated by overexpression of B7-H1 (PDL1). *Human Immunology, 67*(8), 568–578. https://doi.org/10.1016/j.humimm.2006.04.013.

Lin, H., Wei, S., Hurt, E. M., Green, M. D., Zhao, L., Vatan, L., et al. (2018). Host expression of PD-L1 determines efficacy of PD-L1 pathway blockade-mediated tumor regression. *The Journal of Clinical Investigation, 128*(2), 805–815. https://doi.org/10.1172/JCI96113.

Maniotis, A. J., Folberg, R., Hess, A., Seftor, E. A., Gardner, L. M., Pe'er, J., et al. (1999). Vascular channel formation by human melanoma cells in vivo and in vitro: vasculogenic mimicry. *The American Journal of Pathology, 155*, 739–752.

Mazzucchelli, L., Blaser, A., Kappeler, A., Schärli, P., Laissue, J. A., Baggiolini, M., et al. (1999). BCA-1is highly expressed in helicobacter pylori-induced mucosa associated lymphoid tissue and gastric carcinoma. *Journal of Clinical Investigation, 104*(10), R49–R54. https://doi.org/10.1172/JCI7830.

Noman, M. Z., Messai, Y., Muret, J., Hasmim, M., & Chouaib, S. (2014). Crosstalk between CTC, immune system and hypoxic tumor microenvironment. *Cancer Microenvironment,* 7(3), 153–160. https://doi.org/10.1007/s12307-014-0157-3.

Okazaki, T., & Honjo, T. (2007). PD-1 and PD-1 ligands: from discovery to clinical application. *International Immunology, 19*, 813–824.

Oron, T., Colussi, O., Marchetau, E., Pernot, S., Nizard, M., Pointet, A. L., et al. (2015). VEGF-A modulates expression of inhibitory checkpoints on $CD8^+$ T cells in tumors. *Journal of Experimental Medicine, 212*, 139–148.

Peng, F., Xu, Z., Wang, J., Chen, Y., Li, Q., Zuo, Y., et al. (2012). Recombinant human endostatin normalizes tumor vasculature and enhances radiation response in xenografted human nasopharyngeal carcinoma models. *PLoS One,* 7(4). https://doi.org/10.1371/journal.pone.0034646. e34646–e34646.

Pennacchietti, S., Michieli, P., Galluzzo, M., Mazzone, M., Giordano, S., & Comoglio, P. M. (2003). Hypoxia promotes invasive growth by transcriptional activation of the met protooncogene. *Cancer Cell, 3*(4), 347–361. https://doi.org/10.1016/S1535-6108(03)00085-0.

Pietras, K., & Hanahan, D. (2005). A multitargeted, metronomic, and maximum-tolerated dose \chemo-switch\ regimen is antiangiogenic, producing objective responses and survival benefit in a mouse model of cancer. *Journal of Clinical Oncology, 23*(5), 939–952. https://doi.org/10.1200/JCO.2005.07.093.

Postow, M. A., Chesney, J., Pavlick, A. C., Robert, C., Grossmann, K., McDermott, D., et al. (2015). Nivolumab and ipilimumab versus ipilimumab in untreated melanoma. *The New England Journal of Medicine, 372*(21), 2006–2017. https://doi.org/10.1056/NEJMoa1414428.

Powell, M. A., Sill, M. W., Goodfellow, P. J., Benbrook, D. M., Lankes, H. A., Leslie, K. K., et al. (2014). A phase II trial of brivanib in recurrent or persistent endometrial cancer: An NRG oncology/gynecologic oncology group study. *Gynecologic Oncology, 135*(1), 38–43. https://doi.org/10.1016/j.ygyno.2014.07.083.

Presta, L. G., Chen, H., O'Connor, S. J., Chisholm, V., Meng, Y. G., Krummen, L., et al. (1997). Humanization of an anti-vascular endothelial growth factor monoclonal antibody for the therapy of solid tumors and other disorders. *Cancer Research, 57*(20), 4593–4599.

Quintieri, L., Selmy, M., & Indraccolo, S. (2014). Metabolic effects of antiangiogenic drugs in tumors: Therapeutic implications. *Biochemical Pharmacology, 89*(2), 162–170. https://doi.org/10.1016/j.bcp.2014.02.018.

Rahat, M. (2019). Targeting angiogenesis with peptide vaccines. *Frontiers in Immunology, 10*, 1924. https://doi.org/10.3389/fimmu.2019.01924. eCollection 2019.

Relf, M., LeJeune, S., & Scott, T. A. (1997). Expression of the angiogenic factors vascular endothelialcell growth factor, acidic and basic fibroblast growth factor, transforming growth factor beta-1, platelet-derived endothelial cell growth factor, placenta growth

factor, andpleiotrophin in human primary breast cancer and its relation to angiogenesis. *Cancer Research, 57*, 963–969.
Ribatti, D. (2016). Tumor refractoriness to anti-VEGF therapy. *Oncotarget, 7*(29), 46668–46677. https://doi.org/10.18632/oncotarget.8694.
Ribatti, D., Nico, B., & Crivellato, E. (2011). The role of pericytes in angiogenesis. *International Journal of Developmental Biology, 55*(3), 261–268. https://doi.org/10.1387/ijdb.103167dr.
Rigamonti, N., Kadioglu, E., Keklikoglou, I., Rmili, C. W., Leow, C. C., & de Palma, M. (2014). Role of angiopoietin-2 in adaptive tumor resistance to VEGF signaling blockade. *Cell Reports, 8*(3), 696–706. https://doi.org/10.1016/j.celrep.2014.06.059.
Romiti, A., Cox, M. C., Sarcina, I., Di Rocco, R., D'Antonio, C., Barucca, V., et al. (2013). Metronomic chemotherapy for cancer treatment: A decade of clinical studies. *Cancer Chemotherapy and Pharmacology, 72*(1), 13–33. https://doi.org/10.1007/s00280-013-2125-x.
Sandler, A., Gray, R., Perry, M. C., Brahmer, J., Schiller, J. H., Dowlati, A., et al. (2006). Paclitaxel-carboplatin alone or with bevacizumab for non-small-cell lung cancer. *New England Journal of Medicine, 355*(24), 2542–2550. https://doi.org/10.1056/NEJMoa061884.
Schupp, J., Krebs, F. K., Zimmer, N., Trzeciak, E., Schuppan, D., & Tuettenberg, A. (2019). Targeting myeloid cells in the tumor sustaining microenvironment. *Cellular Immunology, 343*. https://doi.org/10.1016/j.cellimm.2017.10.013.
Shacter, E., & Weitzman, S. A. (2002). Chronic inflammation and cancer. *Oncology (Williston Park, N.Y.), 16*(2), 217–232.
Taube, J. M., Anders, R. A., Young, G. D., Xu, H., Sharma, R., McMiller, T. L., et al. (2012). Colocalization of inflammatory response with B7-h1 expression in human melanocytic lesions supports an adaptive resistance mechanism of immune escape. *Science Translational Medicine, 4*(127). https://doi.org/10.1126/scitranslmed.3003689, 127ra37-127ra37.
Teft, W. A., Kirchhof, M. G., & Madrenas, J. (2006). A molecular perspective of CTLA4 function. *Annual Review of Immunology, 34*, 65–97.
Topalian, S. L., Taube, J. M., Anders, R. A., & Pardoll, D. M. (2016). Mechanism-driven biomarkers to guide immune checkpoint blockade in cancer therapy. *Nature Reviews Cancer, 16*(5), 275–287. https://doi.org/10.1038/nrc.2016.36.
Trusolino, L., Bertotti, A., & Comoglio, P. M. (2010). MET signalling: Principles and functions in development, organ regeneration and cancer. *Nature Reviews Molecular Cell Biology, 11*(12), 834–848. https://doi.org/10.1038/nrm3012.
van der Woude, C. J., Kleibeuker, J. H., Jansen, P. L. M., & Moshage, H. (2004). Chronic inflammation, apoptosis and (pre-)malignant lesions in the gastro-intestinal tract. *Apoptosis, 9*(2), 123–130. https://doi.org/10.1023/b:appt.0000018794.26438.22.
Wagner, S. C., Ichim, T. E., Ma, H., Szymanski, J., Perez, J. A., Lopez, J., et al. (2015). Cancer anti-angiogenesis vaccines: Is the tumor vasculature antigenically unique? *Journal of Translational Medicine, 13*(1). https://doi.org/10.1186/s12967-015-0688-5.
Wang, Q., Liu, F., & Liu, L. (2017). Prognostic significance of PD-L1 in solid tumor: An updated meta-analysis. *Medicine (United States), 96*(18). https://doi.org/10.1097/MD.0000000000006369.
Westendorf, A. M., Skibbe, K., Adamczyk, A., Buer, J., Geffers, R., Hansen, W., et al. (2017). Hypoxia enhances immunosuppression by inhibiting CD4+ Effector T cell function and promoting treg activity. *Cellular Physiology and Biochemistry, 41*(4), 1271–1284. https://doi.org/10.1159/000464429.
Whittle, M. C., & Hingorani, S. R. (2019). Fibroblasts in pancreatic ductal adenocarcinoma: Biological mechanisms and therapeutic targets. *Gastroenterology, 156*(7), 2085–2096. https://doi.org/10.1053/j.gastro.2018.12.044.

Wilson, W. R., & Hay, M. P. (2011). Targeting hypoxia in cancer therapy. *Nature Reviews Cancer, 11*(6), 393–410. https://doi.org/10.1038/nrc3064.
Winkler, F., Kozin, S., Tong, R., Chae, S., Booth, M., Garkavtsev, I., et al. (2004). Kinetics of vascular normalization by VEGFR2 blockade governs brain tumor response to radiation role of oxygenation, angiopoietin-1, and matrix metalloproteinases. *Cancer Cell, 6*(6), 553–563. https://doi.org/10.1016/s1535-6108(04)00305-8.
Wu, P., Wu, D., & Li, L. (2015). PDL-1 and survival in solid tumors: A meta-analysis. *PLoS One, 10*.
Yu, P., & Fu, Y. X. (2006). Tumor-infiltrating T lymphocytes: Friends or foes? *Laboratory Investigation, 86*(3), 231–245. https://doi.org/10.1038/labinvest.3700389.

Index

Note: 'Page numbers followed by "b" indicate boxes, those followed by "f" indicate figures and those followed by "t" indicate tables.'

G

H

I

L

M

N